CW01395076

FISHERIES MANAGEMENT IN THE EUROPEAN UNION

Fisheries Management in the European Union

APOSTOLOS KARAGIANNAKOS
Economics Department
University of Aberdeen

Avebury

Aldershot · Brookfield USA · Hong Kong · Singapore · Sydney

© A. Karagiannakos 1995

Published by
Avebury
Ashgate Publishing Limited
Gower House
Croft Road
Aldershot
Hants GU11 3HR
England

Ashgate Publishing Company
Old Post Road
Brookfield
Vermont 05036
USA

A CIP catalogue record for this book is available from the British Library

Library of Congress Catalog Card Number: 94-80259

ISBN 1 85972 070 6

Printed in Great Britain by Antony Rowe Ltd, Chippenham, Wiltshire

Contents

vi

Figures and tables

Acknowledgements

I would like to thank Mr. Robert Shaw for his rigorous attention to detail and his meticulous and patient advice during the writing of this book. I am particularly indebted to Prof. P.J. Sloane who provided the facilities in the Economics Department of Aberdeen University and offered his intellectual support on many occasions. I am also grateful to Mr. Neil MacKellar for the confidence and opportunities he gave me by offering his generous advice and suggestions. Many thanks to Dr Andreas Yannopoulos for his helpful advice and his friendship. My deepest gratitude is due to my parents, Yannis and Evanthia, late grandfather Apostolos, grandmother Egli, my Titi, my wife Vanda and my parents-in-law Vassilis and Voula for their continuous emotional encouragement and inexhaustible support throughout the period of my commitment to this project. Finally, I am also thankful to my daughter Eva for the delightful intervals she offered me by intruding my office.

List of abbreviations

ACFM:	Advisory Committee for Fisheries Management
ACP:	African, Carribean and Pacific
CAP:	Common Agricultural Policy
CCT:	Common Customs Tariff
CECAF:	Fisheries Committee for Eastern Central Atlantic
CFP:	Common Fisheries Policy
COM:	European Community Directive
COREPER:	Committee of Permanent Representatives
CPUE:	Catch per unit effort
Dec:	Decision
DG:	Directorate General
EAGGF:	European Agricultural Guidance and Guarantee Fund
EC:	European Community
ECOSOC:	Economic and Social Committee of the European Communities
ECSC:	European Coal and Steel Community
ECU:	European Currency Unit (same value as EUA)
EDF:	European Development Fund
EEZ:	Exclusive Economic Zone
EFTA:	European Free Trade Association
ERDF:	European Regional Development Fund
ESF:	European Social Fund
EU:	European Union
EUA:	EEC Unit of Account
FAO:	Food and Agriculture Organization
GATT:	General Agreement on Tariffs and Trade
GDP:	Gross Domestic Product
GFCM:	General Fisheries Council for the Mediterranean

GRT:	Gross Registered Tonnes
GSP:	Generalised System of Preferences
IATTC:	International American Tropical Tuna Commission
ISO:	International Organization for Standardization
ITQ:	Individual Transferable Quotas
ICCAT:	International Convention for the Conservation of Atlantic Tunas
ICES:	International Council for the Exploration of the Sea
ICSEAF:	International Commission for South East Atlantic Fisheries
IPHC:	International Pacific Halibut Commission
MAGP:	Multi-Annual Guidance Program
MATAC:	Multi-Annual TAC
MEP:	Members of European Parliament
MEY:	Maximum Economic Yield
MRAP:	Most Rapid Approach Path
MSTAC:	Multi-Species TAC
MSVPA:	Multi-Species Virtual Population Analysis
MSY:	Maximum Sustainable Yield
NEAFC:	North-East Atlantic Fisheries Commission
OECD:	Organization for Economic Co-Operation and Development
OJ:	Official Journal of the European Communities
Reg:	Regulation
SSB:	Spawning Stock Biomass
STCF:	Scientific and Technical Committee for Fisheries
STECF:	Scientific, Technical and Economic Committee for Fisheries
TAC:	Total Allowable Catch
UNCLOS:	United Nations Conferences on the Law of the Sea
UNCTAD:	United Nation. Conference on Trade and Development
VPA:	Virtual Population Analysis

Introduction

The persisting pressure for practical, viable and applicable decisions for managing marine resources is dominant in most fisheries of the world and European Communities (E.C.) fisheries are no exception. Though the European Communities were instituted by the Treaty of Rome in 1957, they did not attempt to establish a Common Fisheries Policy (CFP) until 1971 and the CFP was not organized in a comprehensive mode until 1983. It is now the case that the concept of the European Communities have been replaced by the European Union which was established by the Treaty on European Union, signed on 7 February 1992 and taking effect from 1 January 1993. The Treaty followed the Maastricht agreement of 1991 and emphasizes not just convergence to achieve the European Monetary Union but also a determination to continue the process of creating an even closer union among the peoples of Europe (Britton and Mayes, 1992). However, in this book the term European Communities will be used, since the fisheries management framework which is in existence at present is based in the previous notion of the European Communities.

A common fisheries policy represented an urgent prerequisite for the effective management of the resource, since fish stocks are notorious for showing little respect for state boundaries, a fact emphasized by the Community's target for free fishing access after 2002. However, the current CFP is widely regarded as being a failure, mostly for allowing a decline in fish stocks and for being unable to deal with excess capacity in fishing vessels. The multiparametric nature of fisheries has burdened the management decision-making with extra difficulties of identifying and balancing factors and objectives which many times are conflicting. The purpose of this book is to document the history of the CFP, paying particular attention to the period since 1983, to identify the sources of the failure and to suggest ways of improvement.

Chapters 1 and 2 survey the literature in fisheries economics and fish population dynamics, which should provide the basis for efficient management. For a long time fisheries economists have been involved in developing static and dynamic models of the fishing industry, whose mathematical complications could sometimes hamper efficient management formation. However, it is unlikely that fisheries regulators have seriously considered the economic principles, by which the industry should be regulated as the present overcapacity seems to suggest. Moreover, the contribution of fish population dynamics models, which tend to oversimplify reality, have been consistently overrated. Nevertheless, they do not cease to provide a useful tool for examining and predicting the condition of the stocks.

After the limitations of the above scientific processes have been discussed, the Common Fisheries Policy (CFP) is examined in detail and its different components are analyzed with respect to their contribution to the effective fisheries management policy formation. Chapter 3 describes the modes of decision-making and administration procedures at the overall level of the Community, while Chapter 4 looks at the evolution of the CFP and describes the means by which policy is formed for fisheries. The case of Mediterranean regulation is also presented.

The next three chapters examine the main constituent parts of the CFP. The main objectives of the CFP consist of ensuring the rational development of the production factors, a fair standard of living for the producers, a stable market and the availability of supplies for consumers. The means by which the Community regulates the industry is by technical and conservation measures, which involve mostly catch restrictions, by the common organization of the market and by the Common Structural policy which aims to deal with the problem of overcapacity.

It should be mentioned here that overall, the Community's total commitment to fisheries represents less than 1% of the Community's total expenditure (68,000 million ECUs), which in turn comprises only 1.2% of the GDP of the twelve Member States in 1993 [169,171]. The Community's expenditure on fisheries is extremely small, when compared with that for Agriculture (35,000 million ECU) which represents 49% of the total Community expenditure. The expenditure on fisheries is only some one sixteenth of the Community's market support for butter and less than half the Community's market support for tobacco and wine in the 1993 budget [169,171]. However, despite the relatively low importance of the fisheries sector in the Community's economy, fisheries management is the subject of extensive studies and disputes. This happens, in our opinion for two main reasons. Firstly, the open-access characteristics of the sea resources, which are difficult to divide by nature, can lead to considerable conflict between states, especially when there are limited resources. Secondly, fisheries can represent an important economic activity locally, since there are coastal communities that are dependent on fisheries, with little alternative employment possibilities.

A detailed account of the resource management and conservation policy is presented in chapter 5. Individual aspects, such as access restrictions, total allowable catches (TACs) and quotas are assessed. Particular attention has been devoted to North Sea demersal fisheries, in order to explore the effect of the TAC system on the actual landings.

In chapter 6 attention is turned to the organization of the market for the first-hand sales of fish and to international trade agreements related to fishery products concluded by the Community. The effect of the Common Organization of the Market on price volatility is explored, by using data from the Scottish fisheries. Chapter 7 describes and assesses the stuctural policy, which deals with the contradictory objectives of fleet modernization and construction on the one hand and the reduction of capacity on the other.

Chapter 8 summarizes the failures of the Common Fisheries Policy and states the need for alternative management strategies. A licensing management scheme is then proposed as an alternative to the existing Common Fisheries Policy. The proposed reforms are based on a theoretical bioeconomic model developed by the author. The above restricted access system could provide a viable and economically efficient alternative to current policies, while it could lead to efficient monitoring and therefore effective conservation of the marine resources. The need for reliable and comprehensive data of the situation in the fishing industry underlines the lack of an effective monitoring and surveillance system. Its establishment would be a desirable short-term measure, which will offer the basis for reforming the CFP. The need for 'aggressive' management techniques, enforced by determined politicians and policymakers is underlined.

1 The fundamentals of fisheries economics

The cumulative problems of the fishing industry require a multidisciplinary approach which calls for a comprehensive and sound knowledge of the various determining features of fisheries for the management of the resource. Seen in this perspective the work of fisheries economists is catalytic since E.C. fisheries regulators should, in terms of article 39(1) (page 91) of the Treaty of Rome, seek for efficiency, equity and price stability. Therefore, this chapter attempts to present an overview of the trends followed by fisheries economists in their effort to understand the way the fishing industry operates. Their expertise on the factors which influence the industry could then be used for the formation of effective fisheries management.

In general, the production function of an industry or a firm Q=f(L,K,T,etc) is the relationship between the output (Q) of a good and the inputs required to make that good. These inputs can be labour (L), capital (K), technical progress (T) and others as raw materials (Pearce, 1986). In terms of the fishery, it is a relationship between the effort applied and the fish caught, and it follows that it depends greatly on the reproductive biology of the fish stock. (A fish stock does not necessarily include all the fish of a particular species. More than one biologically independent stock of any species may exist).

A simple biological theory of population dynamics is normally employed, namely the Schaefer logistic analysis, where the growth of the fish stock measured in weight is assumed to be a function of its size and weight (Schaefer, 1954; 1957; 1959). According to this theory, the biomass, x, of the unexploited fish stock will tend to increase at various growth rates depending upon its size. The biomass will grow towards some maximum weight that, once reached, will be maintained. It is assumed that the net increase in the biomass of a fishery as a function of the population, can be depicted as a bell-shaped curve (figure 1.1a) The curve is based on certain assumptions about

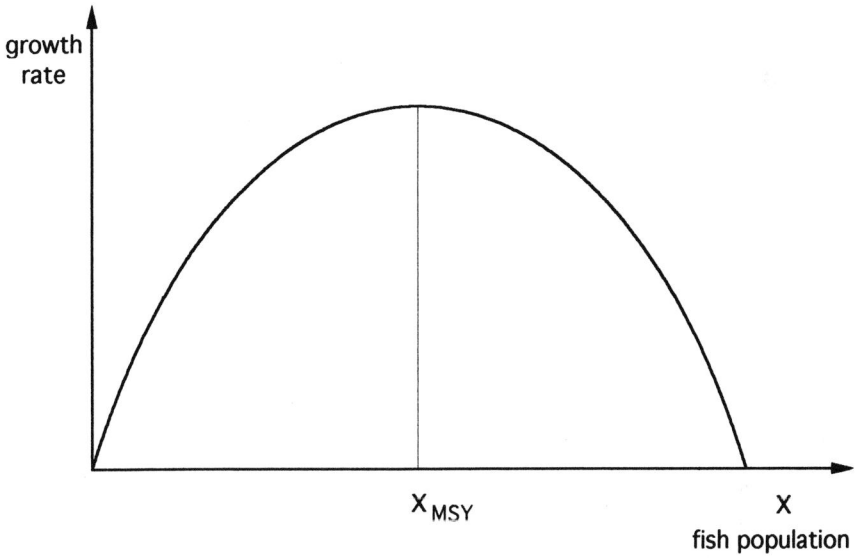

(a)

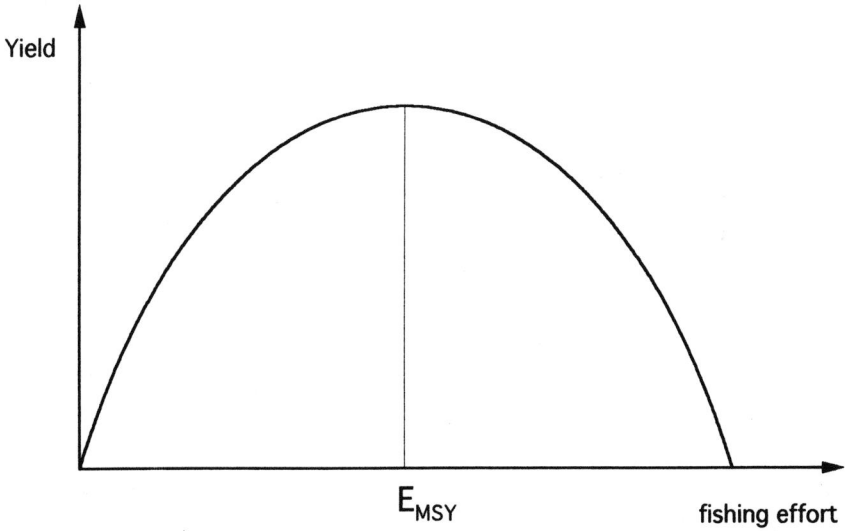

(b)

Figure 1.1   Fish population growth curve (a) and
sustainable yield-effort curve (b)
based on Schaefer logistic model

stock recruitment, individual growth and mortality and also about features such as salinity, temperature and rate of photosynthesis.

When fishing is introduced to this model, it is postulated that the yield, or catch (Y) in any period will depend on only two factors: a) the size of the fish population at the beginning of the period, p and b) the amount of fishing effort, E.

$$Y = f(p, E).$$

For each level of effort, then, there is an equilibrium population size. As long as the curves are as drawn in figure 1.1b, the catch obtained from a level of effort and its corresponding equilibrium population is called a sustainable or a sustained yield.

It is sustainable because the population size will not be affected by fishing, since the catch is replaced by the natural increase over the long run. Therefore, the same level of effort will yield the same catch in any period. The locus of points representing sustainable yield catches for each level of effort is called the sustainable or sustained yield curve (figure 1.1b). The vertical axis measures catch in weight, and the horizontal axis measures effort. The Schaefer yield curve reaches a maximum, referred to as the maximum sustainable yield (MSY), and then declines as effort is further increased. At sufficiently high sustained effort levels, yield falls to zero. Biological overfishing is said to occur whenever sustainable yield falls below MSY. The biological explanation for such an outcome is that intensive exploitation reduces the fish stock to a level at which productivity begins to decline.

Although a number of bioeconomic models have been developed over the years (see below), perhaps the most straightforward and widely used, particularly by economists is the so called Gordon-Schaefer model named after the economist H. S. Gordon and the biologist M. B. Schaefer. Gordon's economic analysis was based upon the simple yield-effort biological curve stated above.

Preliminary analysis: Open-access fishery

The economic theory of an open access or common property fishery was developed by Gordon (1954). Reviewing the literature it is sometimes not clear that there is a distinction between open-access and common property resources. Some scholars, (Ciriacy-Wantrup and Bishop, 1975), argue that the notion of property implies the exclusion of non-owners. Therefore, before stating the bioeconomic Gordon-Schaefer model it is essential for management purposes to make clear that critical distinction. As Bromley (1985) defines it, open-access (*res nullius*) is free for all, while common property (*res communes*) represents a well defined set of institutional arrangements concerning who may make use of a resource and the rules governing how the accepted users shall conduct themselves.

Regarding now Gordon's model based on the parabolic yield-effort curve described above, it can be assumed that it is a static equilibrium model. It should be recalled that each point on this curve corresponds to the sustainable yield Y(E), measured for example in terms of biomass at a given level of fishing effort, E. If we assume a constant price, p, per unit of harvest biomass, then the function

$$TR=pY(E)$$

represents the total sustainable revenue resulting from the effort, E.

This curve shown in figure 1.2 has the same parabolic shape as the yield-effort curve. It must be underlined here that price has been assumed constant. This assumption obviously is contradictory to the nature of an open-access fishery and is highly restrictive; later, it will be assumed variable.

To introduce a total cost curve, TC, in the same figure, it is assumed that, in the simplest case, the costs of fishing are proportional to the effort expended:

$$TC=c \cdot E$$

The difference between total sustained revenue, TR, and total cost, TC, is called the sustainable economic rent provided by the fishery resource at each given level of effort, E.

$$Sustainable\ economic\ rent=TR-TC=pY(E)-cE$$

Gordon's basic argument is that, in an open-access fishery resource, rents will be dissipated over a period of time. Effort tends to reach an equilibrium (bioeconomic equilibrium) at the level E=E* at which total revenue, TR, equals total cost, TC, and thus the economic forces affecting fishermen and the forces of the biological productivity of the resource are in balance.

A simple argument supports the prediction of bioeconomic equilibrium. First, effort clearly cannot exceed the level E* in the long run, because then fishermen's opportunity costs would exceed their revenues and some fishermen -at least- would take up alternative employment. Conversely, in the absence of entry restrictions, a level of effort below E would induce additional fishermen to enter the fishery, attracted by revenues greater than what they can achieve elsewhere. Thus, a stable equilibrium is established at E=E*. Although this is an oversimplification in several important respects, in summary it can be

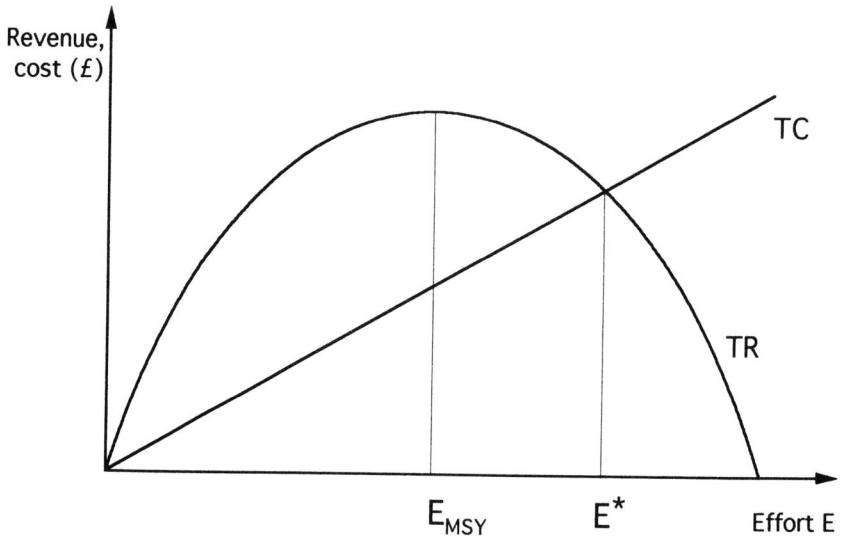

Figure 1.2   The static Gordon model and bioeconomic
equilibrium in an open-access fishery

applied to explain many observable characteristics of the open-access exploitation of natural resources.

Opportunity cost

In an economic analysis of an industry, cost is usually understood in terms of opportunity costs. In some fisheries, however, this may not be the case. Although Gordon (1954) considered the case theoretically, his model did not define the term "cost" Thus, in the simplified analysis above, such things as fuel, bait, and food for the crew, could easily be put to use in other sectors of the economy. Thus, market price would be a good measure of their opportunity cost. However, this is not so, for much of the labour force in the fishing industry.

A related prediction of the Gordon theory is that the fishermen who remain in an open-access fishery tend to have the least number of alternative employment opportunities available (i.e. the lowest opportunity costs). This arises because many fishermen are not equipped for other types of work and additionally they are geographically isolated from the other sources of employment. The incomes received from fishing are usually not a good measure of their opportunity cost but are often just the amount necessary to keep them working. (This will be discussed further below). In addition, the concept of opportunity cost also has a time dimension which is especially relevant in the present context. As Cunningham and his colleagues (1980) suggest, time is in a sense an input into the production function since the harvesting of fish is time-dependent; that is to say, the biomass depends inter alia on the length of time it has been allowed to grow. Additionally, Cunningham and his colleagues (1980) postulate that time, like other inputs, has an opportunity cost because the fishery can be treated not only as a renewable resource that produces a stream of net benefits in perpetuity but as an item that can, in principle, be consumed instantly.

Externalities

An externality is said to occur when a decision variable of one economic agent enters into the welfare or profit function of some other economic agent. Common examples include pollution of air and water. This is an example of an external diseconomy since the consumption of pollution may be presumed to reduce the welfare or profit of other economic agents. Such costs imposed on a large number of people are called external diseconomies. In a competitive market system the myriad decisions of individual consumers and producers are coordinated by means of the set of prices that each of these agents face. As a consequence the effects of decisions of one economic agent upon another are transmitted by the price system.

The pricing of "environmental externalities" for the cost-benefit analysis of a developmental project assert that it involves two steps (Goodland *et al.*, 1989). The first step is the physical estimation of the environmental effects and the second step is their monetary valuation. The contemporary approach to correcting the deficiencies of the price mechanism in economic cost-benefit analysis, when addressing the ecological aspects of development projects and policies, is to assign "shadow prices" to environmental effects. Shadow prices are economic values imputed to goods or services which have no market price, or for which the market price is considered "distorted", so that it does not accurately reflect actual social costs and benefits. The shadow price also represents the opportunity cost of producing or consuming a commodity which is generally traded in the economy. A variety of cost-benefit analysis techniques can be used in such valuations.

For example, the most widely used technique is the Contingent Valuation method (Hanley, 1989) in which a sample of individuals is directly questioned about their willingness-to-pay in order to benefit from welfare-increasing changes in environmental services. It is not the scope of this book to analyse this method; however great controversy exists in the literature about the efficiency of the method to measure the economic value of environmental externalities (Loomis and Walsh, 1986; Stevens *et al.*, 1991). Similar techniques were applied in the wildlands of an Indonesian island for an irrigation project of rice culture and the establishment of a dense tropical forest as a national park (Ledec and Goodland, 1988). In this case, pricing environmental externalities and trading off between irrigation works and protected park area was relatively easy since the forest provided the watershed catchment area. Therefore, conserving the forest would enhance the productivity of the irrigation project. In contrast, the validity of imputing economic values to coastal and marine wildlife by the use of the Contingent Valuation method is disputable, as uncertainty must be considered in such studies (Whitehead, 1993). Moreover, the economic valuation of marine environmental degradation by fishing activities seems a long term goal to be attained. This happens because a thick cloud of uncertainty covers both its biological and environmental aspects as well as the economic analysis of the forces regulating the market in fish.

Possible means of controlling externalities, i.e. taxes or gear restrictions, suffer from many deficiencies and tend systematically to underestimate environmental values. However, they are capable of providing partial "lower bound" estimates of many (though not all) environmental benefits and costs.

The use of taxes to control the common-property externalities of fishery resources was first discussed by Gordon (1954). Subsequent work by Bell (1972), Smith (1969) and Quirk and Smith (1970) have identified at least four distinct externalities:

1. Stock externalities which arise when the cost of catching fish depends on the fish population.
2. Crowding externalities which arise from the congestion of boats and gear on the fishing grounds. Thus the operation of one fishing vessel imposes costs on other vessels because it reduces the fish stock and thereby increases unit harvest costs for all vessels. On the other hand, external benefits can occur when one vessel locates a school of fish and other vessels "home in on it".
3. Interspecies externalities, which arise if the harvest of one species affects the abundance of the other species.
4. Gear or mesh externalities which arise if the choice of gear affects the growth of the fish population.

In conclusion, the complicated nature of the open-access exploitation of fisheries creates a host of problems of externalities and as Clark (1990) confirms, externalities do not seem to lend themselves readily to analytical treatment. However, further knowledge of environmental, biological and economic forces affecting fisheries would contribute to successful and appropriate management techniques for controlling fisheries.

Maximum Economic Yield

Resource rent is defined as a surplus value over and above the opportunity costs for all factors of production. It arises from ownership of or access to a valuable resource in limited supply. Proper use of the fish stock requires that resources be utilised to exploit it, so that the present value of future net returns is maximised, that is, that the stream of net incomes earned, properly discounted, is a maximum. Maximum resource rents are obtained at a certain level of effort E_{MEY} (figure 1.3). In the literature this point is often referred to as the position of maximum economic yield (MEY).

In other words, if costs and revenues have been correctly measured, the proper amount of effort in a static sense will be E_{MEY} (figure 1.3). The marginal revenue of effort (i.e. the change in sustained catch brought about by the change in effort, multiplied by the fixed price of fish) is downward sloping because the marginal catch per unit of effort decreases as greater intensity of

11

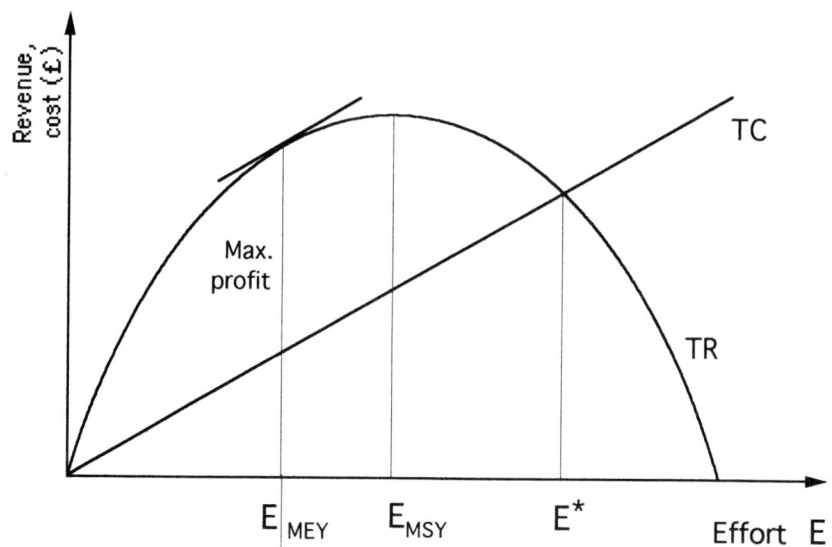

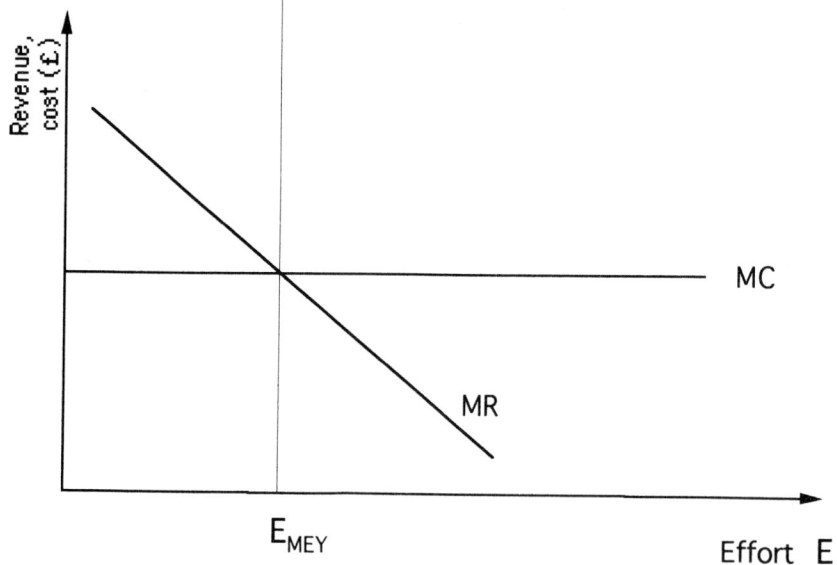

Figure 1.3   Maximization of the sustainable
economic rent

fishing develops and just balances the marginal cost of effort at the point of intersection. The marginal cost curve, that shows the change in cost due to a change in effort is by assumption constant (figure 3). Let us consider a fishery operating at $E>E_{MEY}$. Profits will decrease as costs increase more than revenue (MC>MR); society is losing, because the additional fish are being taken at a cost greater in terms of other output foregone than their value to consumers. On the other hand, if $E<E_{MEY}$ profit will fall again, implying that revenues must be falling faster than cost and that resources are being released for other production, which is valued less than the foregone fish. Thus, in terms of the above, E_{MEY} is the optimal allocation of effort to the fishery, since the value to society of the last unit of fish caught (marginal revenue) just balances the cost of providing it (marginal cost), while the annual profit to the fishery as a whole is maximised. In addition, it must be noted here that at this stage E_{MEY} effort is still below the level of E_{MSY} and biological overfishing does not occur.

However, although it is tempting to suggest the maximum of sustainable economic rent as the criterion for determining the optimal level of fishing effort, this suggestion neglects an essential ingredient of the problem, namely, the dynamics of both economic and biological processes. Indeed, in an open-access fishery the resource rent will accrue to the fishermen and vessel owners and as a consequence they will be earning returns above their opportunity costs. Moreover, additional fishermen and vessels will enter the fishery and finally equilibrium will be achieved only at $E=E^*$ where total revenue TR equals total cost TC (figure 1.3). At this point the economic rent is completely "dissipated".

The same conclusion can be elicited from the figure 1.4, where total costs and total revenues have been plotted against biomass, x. Once biological overfishing occurs and x falls below x_{MSY}, further decreases in the stock level result in decreased TR and increased TC. It is obvious therefore, that the traditional (since the end of the Second World War) management criterion of "full utilization" (Larkin, 1977) of the resource, will lead to economic overexploitation of the resource. Alternatively, the MSY criterion should be rejected since it ignores the costs of harvesting and the true nature of the benefits derived from fishing.

One aspect of the problem that does not show up in Gordon's static model is this: it is never possible to move instantaneously from one bioeconomic equilibrium to an improved position such as MEY. Such a shift requires that the fish be rehabilitated to more productive levels; stock rehabilitation always takes time to accomplish, depending much on the fish reproductive cycles. For example, the reproductive cycle of the depleted Pacific salmon stock is three to five years (Pearse, 1979). Thus, reductions in current harvest levels only lead

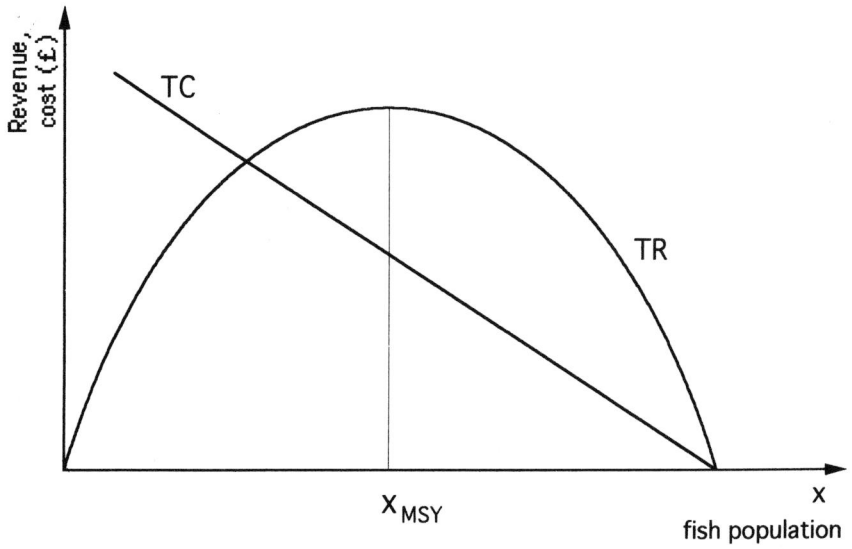

Figure 1.4   Cost and revenue in terms of the
               biomass level X, (Clark, 1985)

to increased recruitment after several years delay. In other words, a reduction in fishing effort will initially result in a reduction in catches and fishermen's revenues, rather than in increased catches and revenues, as might be suggested by Gordon's simple model. The fundamental problem then becomes one of determining the optimal trade-off to be made between current and future harvests. Although this problem is the very essence of resource conservation, it is an exceedingly difficult one from a political and ethical viewpoint.

The standard device used to handle questions of intertemporal economic benefits is time discounting. Given the objective of the maximization of sustained economic rent, and the supposition that society is, or should be, willing to make arbitrary current sacrifices to benefit future generations, some authors (Clark, 1990; Page, 1976) have argued that consideration of intergenerational equity requires the use of a zero discount rate in public conservation decisions. Others have recommended 2-3% per annum as appropriate values for the social rate of discount. Such rates reflect the real long term rate of interest on secure investments, such as bonds (Feldstein, 1964). Private real discount rates have been estimated at 10% per annum for large firms to as high as 39% p.a. for individuals (Hausman, 1979). The whole notion of resource conservation, based on the underlying assumption that current sacrifices are immaterial if they ultimately lead to a permanent increase in economic benefits, has long been known and discussed by biologists and economists. Nevertheless, contemporary fisheries history proves that it is easier and more trivial to postulate theoretical ideals than to apply them.

By way of conclusion from the above discussion it can be asserted that Gordon's bioeconomic model, although successful in explaining the overfishing phenomenon, appears inadequate as a model for an economically desirable management policy. Ironically, as the static model was achieving some measure of recognition among policy makers, it was increasingly viewed sceptically by economists (Munro and Scott, 1985). This static equilibrium model is too simplistic to provide reasonable predictions of the behaviour of resource exploiters or to serve as a reliable guide to a socially desirable management policy.

Further analysis: dynamic fisheries models

As any attempt at a management policy for fisheries is really a question of the extent to which it is appropriate for society to invest/disinvest in the resource, it becomes apparent that a capital theoretic approach is called for. Scott's (1955) article was a first pioneering attempt to cast the problem of fishery resource management as an aspect of capital theory, using Gordon's model in a dynamic framework. Although Gordon in his 1956 paper (Gordon, 1956) writes forcefully of the need for a capital theoretic approach to fisheries economics, and in early 1960s Crutchfield and Zellner (1962) formulated the problem in terms of a dynamic mathematical model, static analysis has remained dominant throughout.

It is beyond the scope of this work to reveal the reasons for that retreat to non-dynamic analysis but it may be mentioned here that Dorfman (1969) has argued that capital theory itself has suffered from an inadequacy of mathematical techniques; since its original formulation by Ramsey (1928) the calculus of variations has been the only mathematical approach to any problem of capital theory. The elimination of the inadequacies of the classical techniques has to a large extent been due to the development of optimal control theory (Bellman, 1957; Pontryagin *et al.*, 1962) that was rapidly applied to modern capital theory. It is because of this that Dorfman (1969) enthusiastically postulates that modern capital theory owes its origins to the development of optimal control theory. After that, it was only a matter of time before it came to be applied in fisheries economics.

The formulation of a mathematical model in fisheries management. The Production Function

A simple model used widely in fisheries economics (Clark, 1973; Crutchfield and Zellner, 1962; Plourde, 1971), has been developed by Clark and Munro (1975) and Clark (1990). It rests upon Schaefer's (1954; 1957) standard general production fishery model, described above. The biological component of this dynamic model is based on the logistic population model:

$$\frac{dx}{dt} = \dot{x} = F(x(t)) = rx(t)\left(1 - \frac{x(t)}{k}\right)$$

where r is a positive constant called the intrinsic growth rate. If $0 < r \leq 1$, the population steadily approaches k without overshooting (May, 1975), resulting in the ogive logistic growth curve (figure 1.5).

In order to model the effect of fishing on population dynamics Schaefer (1954) altered the equation to:

$$dx/dt = F(x(t)) - h(t)$$

where x(t) denotes the biomass at time t and h(t) the harvest rate at time t. It is assumed that:

16

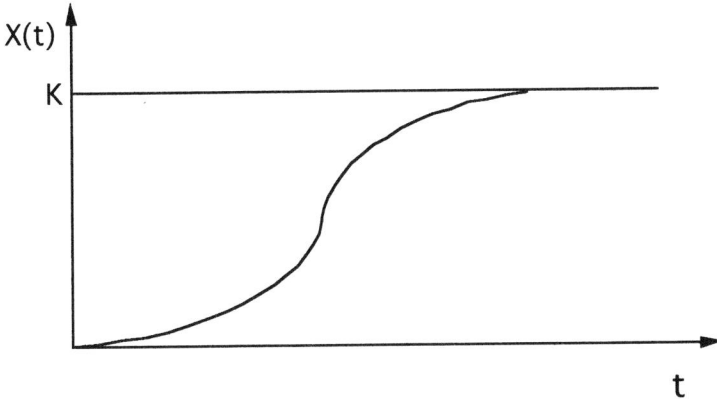

Figure 1.5 The ogive logistic growth curve when 0<r≤1

$F(x)>0$ for $0<x<k$, $F(0)=F(k)=0$ and $F''(x)<0$ for all $x>0$,

where K is called the environmental carrying capacity and denotes the natural equilibrium level of x; i.e.:

if $h(t)=0$ then $\lim_{t \to \infty} x(t)=k$

For a renewable resource, such as fish stocks, it is usually assumed that the rate of harvest (or yield) per unit time is a function of: a) the economic inputs devoted to harvesting and b) the available stock. A sustainable aggregate measure for the various economic inputs (labour plus capital services devoted to harvesting) is referred to as "effort" and denoted by E(t). In general, the production function may be written explicitly as:

$$Y(t)=H(E(t),x(t))$$

where Y(t) is the rate of harvest, measured in the same units as the resource stock x(t).

Clark (1985), advances some basic definitions of effort and relates them to fishing mortality. These definitions are not all necessarily intended to be directly operational. A production function commonly used in fishery management is derived from two assumptions (Clark, 1985):

a) catch per unit effort Y/E is directly proportional to the density of fish in the sea
b)   the density of fish is directly proportional to the abundance x(t).

17

This production function is :

$$Y(t)=qE(t)x(t),$$

where q is the catchability coefficient and is a constant.

When these assumptions are thought to be unrealistic, a more general form of production function may be preferable (Clark *et al.*, 1986), such as:

$$Y(t)=qE^a(t)x^b(t).$$

The above algebraic form is based on the Cobb-Douglas production function. The variables x(t) and E(t) are subject to the constraints $x(t) \geq 0$ and $E(t) \geq 0$, a and b are constants , and for the sake of linearity this model (Clark *et al.*, 1986) assumes a=1 and $b \geq 0$.

Equilibrium supply curve based on Schaefer Model

Considering the Schaefer model in the form,

$$\frac{dx}{dt} = \dot{x} = F(x(t)) = rx(t)\left(1 - \frac{x(t)}{k}\right) - qE(t)x(t)$$

and the flow of the sustainable economic rent that is equal to

Sustainable economic rent=TR-TC=pY(E)-cE=(pqx(t) - c)E

the conditions for a bionomic equilibrium in the open access fishery are as follows

$$Y(t)= qE(t)x(t) = rx(t)\left(1 - \frac{x(t)}{k}\right) \quad \text{and}$$

$$pqx(t) - c = 0$$

Thus, expressing the sustained yield in terms of the price, p we derive

$$Y = \frac{rc}{pq}\left(1 - \frac{c}{pqk}\right).$$

The graphical representation of this curve, (S) in figure 1.6, is the equilibrium curve for the open-access fishery based on the Schaefer model. This curve represents a long term supply curve, since it relates the long term equilibrium output of the industry to each given price. This type of supply curve is called for obvious reasons backward bending. A typical backward-sloping supply curve usually (contrary to the above curve for the fishing industry) is a short-run labour supply curve where changes along this "supply curve" are

18

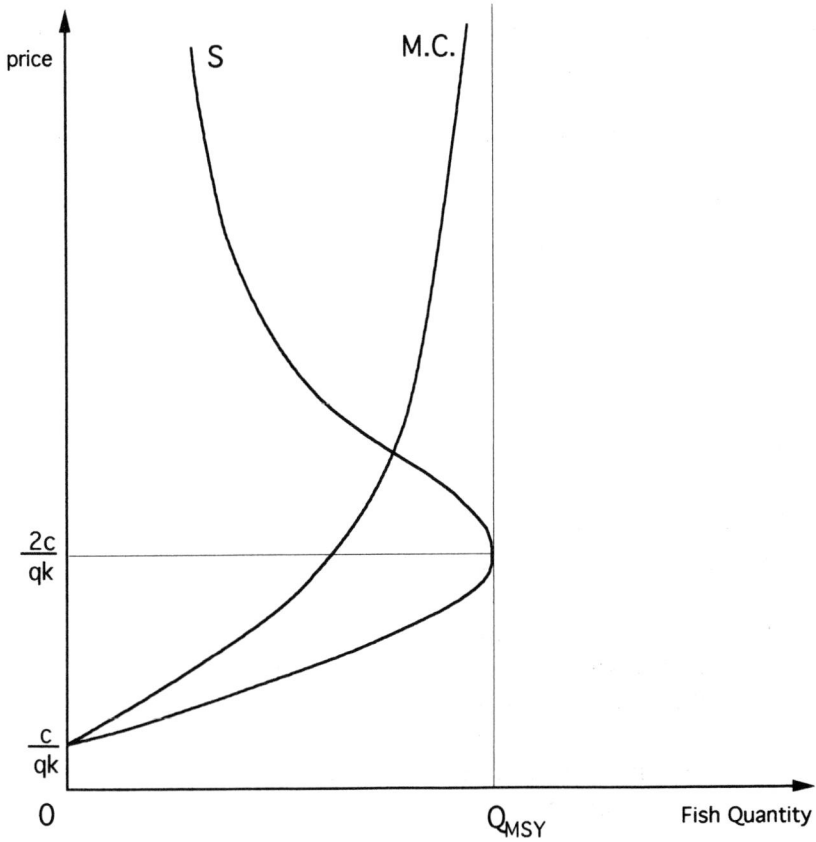

Figure 1.6   The long-term backward-bending
supply and marginal cost curve of the
open-access fishery

irreversible. However, as Copes (1970) asserts, the origin of the backward bending curve for the fishing industry is quite different. Because of such peculiarities the long-run backward-sloping curve for the fishing industry is rather an average cost curve than the supply curve for the industry. The marginal cost curve (MC) that is derived from the backward sloping "supply curve" for the fishing industry (Copes, 1970) is shown in figure 1.6.

When the price of fish, p is $p<c/qk$ the fishery is closed and there is no economic reason for catching fish. When a stock is exploited at the biologically optimum exploitation rate of MSY the catch reaches its maximum level and the price of fish reaches $p=2c/qk$. When fishing effort exceeds the level of E_{MSY} the yield decreases toward zero as the price $p \rightarrow +\infty$. This backward-bending supply shows that when biological overfishing takes place the price unit for fish is higher as the fish stock is driven to depletion.

Discounting

With regard to the above preliminary analysis, it can be postulated here that optimal fishing in terms of the maximization of economic yield is characterized in the static Gordon-Schaefer model by the following objective: maximize the flow of economic rent $\pi(x,E)$ subject to $dx/dt=0$ in order for the fish population to remain constant.

Alternatively, it has been recognized that the capital theory approach to resource stocks makes it essential to define the value of a capital asset (such as fish stocks) as equal to the present value of net future revenues that it is expected to yield. From this viewpoint, resource management simply becomes a special problem in capital theory. Discounting is a technique for calculating the present value of a future stream of net incomes. In discrete time models, with a sequence of payments N_0, N_1,..., N_T due in years t=0,1,...,T the present value of future incomes N_t is calculated according to:

$$N = \sum_{0}^{T} \frac{N_t}{(1+\delta)^t} = \sum_{0}^{T} \rho^t N_t$$

where $\rho=1/(1+\delta)$ is the discount factor and δ is the periodic discount rate.

In continuous time, the present value of net incomes N(t), $0 \le t \le T$ may be calculated according to:

$$N = \int_{0}^{T} N(t) e^{-rt} dt$$

In this equation the term e^{-rt} is the continuous (or instantaneous) discount factor and r is the instantaneous rate of discount. If the time units of the above equations are the same it can be assumed that:

$$e^{-r} = \frac{1}{1+\delta} \quad \text{i.e.} \quad r = \ln(1+\delta)$$

For varying discount rates the following formulae can be applied for the discount factor:

$$\text{discrete time} \quad \rho(t) = \frac{1}{1+\delta_1} \cdot \frac{1}{1+\delta_2} \cdots \frac{1}{1+\delta_t}$$

$$\text{continuous time} \quad \rho(t) = \exp(-\int_0^t r(s) \, ds)$$

When discounting is now applied to the dynamic models so that the "best" dynamic allocation is achieved, our objective is represented by functions which maximize the present value of the flow of net social benefits (or utility, U(t)). Therefore, with a discount rate $\delta > 0$ and solving for the time paths of, say, y(t) and x(t) this may be expressed as:

$$\max \int_0^t U(y(t),x(t)) \, e^{-\delta t} \, dt$$

As for the discrete time analogue with the discount factor $\rho = 1/1 + \delta$ this is:

$$\max. \sum_0^T \rho^t U(y_t,x_t)$$

In its most general statement, it is assumed that society obtains net benefits, U, from the fish resource which depends upon the harvest Y(t) and upon the stock (or population) itself X(t), since a larger stock lowers the cost of a given level of harvest.

Thus, the optimization problem for the dynamic fisheries model becomes:

$$\max. \int_0^t U(Y(t),X(t)) \, e^{-\delta t} \, dt,$$

$$\text{subject to} \quad \frac{dX}{dt} = \dot{x} = F(X(t)) - H(X(t),E(t)), \text{ for } X(0) = X_0,$$

by replacing the *a priori* constant $\dot{x} = 0$ used in the static model with the change in the resource stock to reflect growth and harvest.

21

The discrete-time analogue is:

$$\text{max.} \sum_0^T \rho^t \, [U(Y_t, x_t)]$$

subject to $x_{t+1} - x_t = F(x_t) - H(x_t, E_t)$, x_0 given.

Let us now reintroduce in mathematical formulation the concepts of "price" and "cost" described in the preliminary analysis shown in figure 1.3 (page 12). By denoting p the per unit price of the resource after harvest and c the per unit cost of effort, gross revenue will be equal to the per unit price times the rate of harvest:

$$R(t)=pY(t)$$

and total cost will be the per unit cost of effort times the amount of effort employed in harvesting the resource at instant t:

$$C(t)=cE(t).$$

Net benefits can be written as U=R-C, i.e.:

$$U(Y,X)=pY(t)-cE(t),$$

and using the harvest production function Y=H(X,E), they can be expressed as:

$$U(Y,X)=pH(X,E)-cE.$$

Therefore, in conclusion, the optimization of net social benefits would be achieved by maximizing the net social benefits subject to the appropriate size of the resource in time, i.e.:

$$\text{max} \int_0^T [pH\,[x(t),E(t)]-cE(t)]\, e^{-\delta t}\, dt$$

subject to $\dot{x} = F[x(t)]-H[x(t),E(t)]$, $x(0)=x_0$

and the discrete-time analogue is:

$$\text{max} \sum_0^T \rho^t \, [pH(x_t,E_t)-cE_t]$$

subject to $x_{t+1}-x_t=F(x_t)-H(x_t,E_t)$, x_0 given.

Optimal control theory and its application to fisheries

The above mathematical model is the starting point for describing the powerful optimization techniques that are based on the calculus of variations. These can deal successfully with a variety of multidimensional and non linear problems and have the title "optimal control theory". Alternatively, optimal control theory can be seen as reformulating the economic theory of fisheries in dynamic terms. It provides an amalgamated result, in a systematic and rigorous manner, by combining the forces of modern capital theory with the biological characteristics of fish stocks. The decisive shift away from static to dynamic analysis, via optimal control theory due to the pioneering work of Plourde (1971; 1970) and Quirk and Smith (1970) has placed fisheries management in a position to analyze effectively what might be termed "adjustment or disequilibrium phases" of fishery management problems.

Furthermore, the techniques of optimal control theory (based initially on Lagrange multipliers, subsequently developed by Kuhn and Tucker, and reaching its acme in L.S. Pontryagin's famous maximum principle (Pontryagin et al., 1962)) are not only tools to attain optimality, but they also possess a clear, theoretical, economic interpretation expressing fundamental economic concepts and ideas in mathematical formulation. However, much effort is required of a reader even with a theoretical economics background if he or she is not to lose sight of the economics by becoming enmeshed in complex mathematical formulations. Since the powerful tools of optimal control theory will be used later in order to explore the potential of measures that aim to reform the CFP it would be appropriate to describe its development in detail.

The development of optimization techniques

In order to demonstrate the development of optimization techniques, a constrained problem of fisheries management will be considered, in a more generic form. In that way, a step by step analysis, with the same main model from beginning to end will be worked out as the main pivot to elucidate that development.

In its most general statement, U can be perceived as the utility function of fisheries and G as the function to model the change in the resource stock resulting from growth and harvest. Thus, the starting constrained problem has the form:

$$\text{maximize } U(x,y)$$

$$\text{subject to } G(F(x),Y)=c$$

or for the sake of simplicity $G(x,y)=c$. x and y are called the "decision variables" and could reflect the fish stock and the harvest correspondingly. c could be a known constant or multiple equality constraints, depending upon the different values of a decision variable. The first order necessary conditions are:

$$\frac{\partial U}{\partial x} = 0 \qquad \text{or} \qquad U_x = 0$$

$$\frac{\partial U}{\partial y} = 0 \qquad \text{or} \qquad U_y = 0$$

and differentiating the constraint equation implies,

$$\frac{\partial G}{\partial x} = 0 \qquad \text{or} \qquad G_x = 0$$

$$\frac{\partial G}{\partial y} = 0 \qquad \text{or} \qquad G_y = 0$$

so that the above equations can be written in the form:

$$U_x - G_x = 0 \quad \text{and} \quad U_y - G_y = 0$$

The above equations could be obtained from a geometrical derivation. In fact, in the x, y plane the growth function $G(x,y)$ can be drawn according to the inverted U-shaped logistic growth model of Schaefer (figure 1.7). In addition, on the same plane, a set of indifference curves can be drawn to represent utility or the net benefits obtained by society from the fish resource.

The indifference map depicts the main properties of the indifference curves which are based on the psychology of the consumer (Baumol, 1977; Laidler and Estrin, 1989). Assume a set of well-behaved social indifference curves, obeying the rules applicable to the indifference curves of an individual, i.e. each indifference curve is convex to the origin with a negative slope where its absolute value diminishes towards the right of the curve. Indifference curves can never meet or intersect, so that only one indifference curve will pass through any one point on the map, and as we move upwards and to the right the indifference curves represent higher and higher levels of utility.

For the present geometrical analysis of the generic constrained problem there is no need to be able to measure utility cardinally. Ordinal utility is sufficient to show diagrammatically the ranking of the benefits obtained by the society from the fish resource in different stages of exploitation.

$E^* = (x^*, y^*)$ is the point on the constrained function $G(x,y)$ for which $U(x,y)$ attains its maximum. This is the utility curve $U^*(x^*, y^*)$, which must be tangent to the growth function $G(x,y)$. If it were not, it would either cut $G(x,y)$ or fail to touch it at all. As for the former case, U^* is greater than any other U below

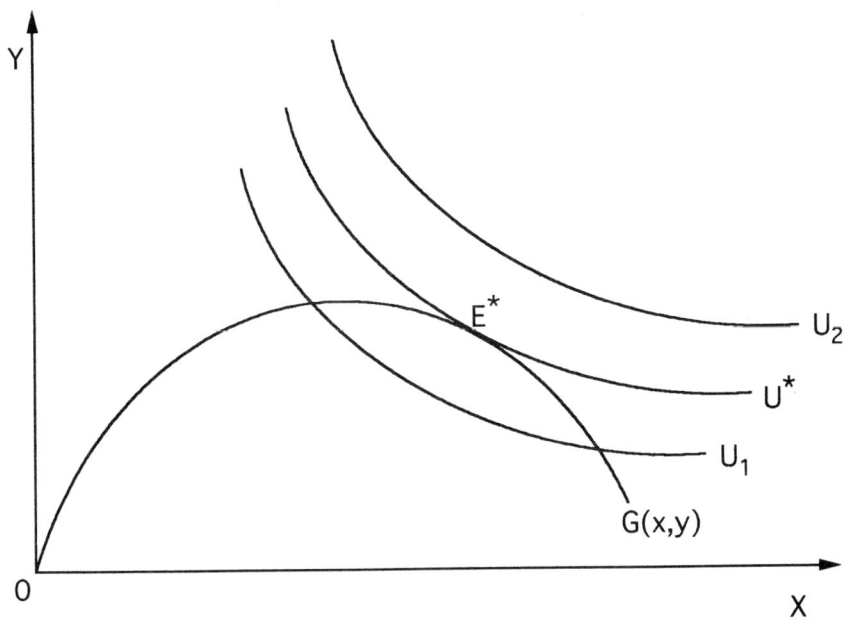

Figure 1.7   Geometrical derivation of the tangency
criterion for the maximization of U(x,y)
at E* given the constraint G(x,y)

U* because of the properties of utility curves mentioned above. As for the latter, the constraint is not satisfied (figure 1.7).

Recalling from the calculus, there is a gradient vector,

$$\vec{\nabla} U = (U_x, U_y)$$

which is normal, that is, constant at any given point and always perpendicular to that point on $U(x,y)$. Because the two curves, $G(x,y)$ and $U(x,y)$, are tangent at E*, their gradient vectors have the same direction, that is

$$\vec{\nabla} U = \lambda \vec{\nabla} G, \text{ for some } \lambda \neq 0.$$

This vector equation implies:

$$U_x = \lambda G_x \quad \text{and} \quad U_y = \lambda G_y \text{ at } E^*.$$

Rewriting the above equations,

$$U_x - \lambda G_x = 0 \text{ and } U_y - \lambda G_y = 0$$

and defining the expression

$$L = U(x,y) - \lambda [G(x,y) - c]$$

the following conditions for a maximum can be obtained:

$$L_x = U_x - \lambda G_x = 0 \text{ and } L_y = U_y - \lambda G_y = 0.$$

The expression L is called the Lagrangian and the number λ is referred to as a Lagrangian multiplier.

Therefore, the use of Lagrangian multipliers leads to the same first order necessary conditions as are obtained by the simple "constraint substitution" technique.

If the first equality constraint is replaced by an inequality we have the problem:

$$\text{maximize } U(x,y)$$

$$\text{subject to } G(x,y) \leq c$$

Because of the inequality constraint, we have a constraint set which consists of all points lying either on the whole $G(x,y)$ curve or on one particular side of the curve. Thus, there are two possibilities:

a)   the optimizing point lies on one side of $G(x,y)$ satisfying the strict inequality $G(x,y) < c$ and $U_x = U_y = 0$ or

b) the optimizing point lies on $G(x,y)$ satisfying the equality $G(x,y) = c$ where the Lagrangian conditions apply and $Lx = Ly = 0$.

Both cases can be combined by a necessary condition generally known as the "Kuhn-Tucker condition" (Intriligator, 1971), that is:

$$Lx = Ly = 0$$

$$\lambda = \begin{cases} =0 \text{ if } G(x,y) < c \\ \geq 0 \text{ if } G(x,y) = 0 \end{cases}$$

For the present analysis there is no possibility of $\lambda = 0$ when $G(x,y) = 0$, because if $\lambda = 0$ the max. of $U(x,y)$ on $G(x,y)$ is also a local or global maximum of $U(x,y)$, but this is not true assuming the properties of $U(x,y)$ mentioned above.

Furthermore, in many problems of constraint optimization such as the problem of fisheries, the decision variables have to be non negative. For these applications the Kuhn-Tucker conditions become:

$$x (Lx) = y (Ly) = 0$$
$$x > 0 \qquad y > 0$$

$$\lambda [G(x,y) - c] = 0 \qquad \lambda \geq 0.$$

Lagrange Multipliers: An economic interpretation

The generic model used for the introduction to control theory is written in such a form that the parameter, time, is obscured. However, time is always related to the condition of the population of the fish stock and also the harvest. Thus, the generic model can be rewritten as:

$$\text{maximize } U(x(t),y(t))$$

$$\text{subject to } G(x(t),y(t)) = c$$

Then, the Lagrangian would be:

$$L = \int_0^T U(x(t) , y(t)) - \lambda(t)[G(x(t),y(t)) - c]$$

with the first order necessary conditions:

$$\frac{\partial L}{\partial x(t)} = \frac{\partial U}{\partial x(t)} + \lambda(t)\frac{\partial G}{\partial x(t)} = 0$$

$$\frac{\partial L}{\partial y(t)} = \frac{\partial U}{\partial y(t)} + \lambda(t)\frac{\partial G}{\partial y(t)} = 0 \qquad (EQ.\ 1)$$

$$\frac{\partial L}{\partial \lambda(t)} = G(x(t),y(t)) - c = 0$$

Utility, U, in the generic model, could also represent the benefit to society from the capital of the resource stock, x, at time t. Thus, $U(x(t),y(t))$ could be redefined in an optimal return scheme as a pay-off function representing net economic return in a period T. From equation (1), when $y(t)$ is required to satisfy the condition, then, in many resource economic problems, $\partial U/\partial y(t)$ is interpreted as the net marginal benefit in period t (Conrad and Clark, 1987).

The second term of the same equation (eq. 1) reflects the influence of harvesting, $y(t)$, on the stock population, $x(t)$. If the stock level is reduced because of an increase in harvesting at time, t, the benefit of the capital of the resource, U, would be also reduced by $\lambda(t)\partial G/\partial y(t)$, since $\partial U/\partial y(t)=\lambda(t)\partial G/\partial y(t)$. J.M. Keynes coined the term "marginal user cost" for the loss in value when a capital asset is reduced by one marginal unit. Therefore, the second term of the equation (1) reflects an intertemporal cost, referred to as "user cost", along the optimal path of $x^*(t)$ (Clark, 1990; Conrad and Clark, 1987).

It could be mentioned here that, for the present model, $\lambda(t)$ represents the shadow price of the resource stock. In other words, $\lambda(t)$ must optimally change through time to reflect the value of the fish imputed: i) from future productivity, ii) from future environmental quality and iii) from the future rate of harvest. Thus, $\lambda(t)$ would be intuitively interpreted as a shadow price (Clark, 1990) or at least the marginal value of the above fishery assets at time t. The interpretation of the Lagrange multiplier as a shadow price is well established in economic analysis (Intriligator, 1971).

The Maximum Principle

With respect to a fisheries optimization problem, we can maximize the net benefit to society $U(x(t), y(t))$ from the capital of the resource stock $x(t)$, subject to a "dynamical system" similar to the dynamic Shaefer fisheries model (page 16):

$$\text{max. } U = \int_0^T U(x(t),y(t),t)dt + K(x(T))$$

$$\text{subject to } \dot{x} = G(x(t),y(t)), \quad \text{for } x(0)=a$$

28

For concreteness, but with some loss of generality, the above parameters of the model could be altered and refined in such a way that they represent a state closer to reality. Therefore, K(T) could represent a final function indicating the value of alternative levels of x(t) at terminal time T, and U(·) could represent net economic return in period t with x(t) continuous and y(t) piecewise continuous.

The Lagrangian expression for the problem is:

$$L = \int_0^t [U(\cdot) + \lambda(t)[G(\cdot) - \dot{x}]]dt + K(x(T))$$

The term $-\lambda(t)\dot{x}$ may be integrated by parts, with the result

$$\int_0^t \dot{\lambda}x(t)dt - [\lambda(T)x(T) - \lambda(0)x(0)].$$

The Lagrangian expression becomes:

$$L = \int_0^t [U(x(t),y(t)) + \lambda(t)G(x(t),y(t)) + \dot{\lambda}x(t)] \, dt + K(x(t)) -$$

$$[\lambda(T)x(T) - \lambda(0)x(0)].$$

At this stage, there can be defined a function, called the "continuous time Hamiltonian", as follows:

$$\mathbb{H}(x(t), y(t), \lambda(t), t) = U(x(t), y(t), t) + \lambda(t)\, G(x(t), y(t))$$

The Hamiltonian function is defined generally as the sum of the intermediate functions of the objective function, U(·), plus the product of the costate, $\lambda(t)$, and the part of the constraint function, G(·), that defines the change rate of the state variable, x(t). Thus, the Lagrangian expression can be rewritten as:

$$L = \int_0^t [\mathbb{H}(x(t), y(t), \lambda(t), t) + \dot{\lambda}x(t)] \, dt + K(\cdot) - [\lambda(t)x(t) - \lambda(0)x(0)].$$

It has been shown previously that, to attain a maximum, all first derivatives in the Lagrangian must be zero. Then, the first order necessary conditions become:

29

$$\frac{\partial L}{\partial y(t)} = \frac{\partial \mathbb{H}(\cdot)}{\partial y(t)} = 0$$

$$\frac{\partial L}{\partial x(t)} = \frac{\partial \mathbb{H}(\cdot)}{\partial x(t)} + \dot{\lambda} = 0, \text{ which implies}$$

$$\dot{\lambda} = -\frac{\partial \mathbb{H}(\cdot)}{\partial x(t)}, \text{ and}$$

$$\frac{\partial L}{\partial x(T)} = -\lambda(T) + K'(\cdot) = 0, \text{ implying } K'(\cdot) = \lambda(T).$$

Thus, as can be seen from the above, the first order conditions can be derived directly as partials of the Hamiltonian. If a terminal time, T, is not specified, then the terminal values of x(T) and λ(T) will through time tend to approach particular values and can, therefore, be treated as constants. As $\partial L/\partial T = 0$, then

$$\frac{\partial L}{\partial T} = \mathbb{H}(T) = \lambda(T)\dot{x}(T) + K'(\cdot)\dot{/}(T) - \dot{\lambda}(T)x(T) - \lambda(T)\dot{/}(T) =$$

$$\mathbb{H}(T) = 0.$$

When $\mathbb{H}(T) = 0$ and $\lambda(T) = K'(\cdot)$ apply, free terminal-value problems arise, and the conditions $\mathbb{H}(T) = 0$, $\lambda(T) = K'(\cdot)$ are referred to as the transversality conditions (Conrad and Clark, 1987; Intriligator, 1971).

Before stating the maximum principle, the terminology of optimal control theory as stated in the standard works (Pontryagin *et al.*, 1962; Lee and Markus, 1968) is set out. Thus:

U(·) is called the objective function.

x(t) is called the state variable.

λ(t) is called the adjoint or costate variable.

$\dot{x} = G(\cdot)$ is called the state equation or equation of motion.

$\partial \mathbb{H}(\cdot)/\partial y(t) = 0$ is called the maximum condition.

$\dot{\lambda} = -\dfrac{\partial \mathbb{H}}{\partial x(t)}$ is called the adjoint equation.

The necessary conditions shown above, namely:

$$\frac{\partial \mathbb{H}(\cdot)}{\partial y(t)} = 0, \quad \dot{\lambda} = -\frac{\partial \mathbb{H}(\cdot)}{\partial y(t)}, \quad \lambda(t) = K'(\cdot) \qquad \text{(EQ. 2)}$$

along with $\mathbb{H}(x(T),y(T),\lambda(T),T) = 0$ are referred to as the Maximum Principle.

The necessary condition $\partial \mathbb{H}/\partial y(t) = 0$ states that the Hamiltonian function is maximized by choice of the control variable y(t). It is valid where an interior maximum exists within the control set along the optimal trajectory (Figure 1.8a). More generally, if there are restrictions on the values taken by the control variable, the maximum condition proved by Pontryagin and his colleagues (1962) becomes:

$$\max \mathbb{H}(x, y, \lambda, t) \quad y \in y(t) \qquad \text{for all } t, t_0 \leq t \leq T.$$

Thus, at any time t, $t_0 \leq t \leq T$, there is either an interior maximum solution at which $\partial \mathbb{H}(\cdot)/\partial y(t) = 0$ (Figure 1.8a), or a boundary solution which is described by the more general maximum condition (Figure 1.8b). Maximum principle is usually taken to imply the more general maximum condition.

Hamiltonian. An economic interpretation

Regarding the Hamiltonian:

$$\mathbb{H}(t) = \mathbb{H}(x(t), \lambda(t), y(t), t) = U(\cdot) + \lambda(t)G(\cdot)$$

the two terms on the right side of the expression can be recognized as value flows. Specifically, $U(\cdot)$ is the flow of net economic returns at instant t, while $G(\cdot)$ is the direct gain at t, say the current flow of investment or else accumulated dividends from fishery assets. In order to express this investment flow in value terms through the change in the state variable, it must be multiplied by the shadow price of capital $\lambda(t)$. Therefore, $\lambda(t)G(\cdot)$ is the increase in the value of the stock x(t) (Conrad and Clark, 1987; Kamien and Schwartz, 1981).

Consequently, the Hamiltonian $H(x, y, t, \lambda)$ represents a performance indicator of the objective function subjected to constraint(s) and gives a "snapshot" idea at a point in time. Thus, it depicts the total rate of increase of total assets. This interpretation implies that the Maximum Principle asserts that an optimal control analysis must maximize the total rate of increase of total assets (Clark, 1990; Kamien and Schwartz, 1981). However, it is essential to know the shadow price $\lambda(t)$ before the optimal choice of y(t) can be made. Indeed, it could be asserted that the Maximum Principle reduces the optimal control problem to a problem determining $\lambda(t)$. However, it is well known that the determination of $\lambda(t)$ by analytic methods is often extremely difficult and in many cases impossible (Clark, 1990).

If we apply discounting (introduced on page 20) in the continuous time model, we would have the following problem:

31

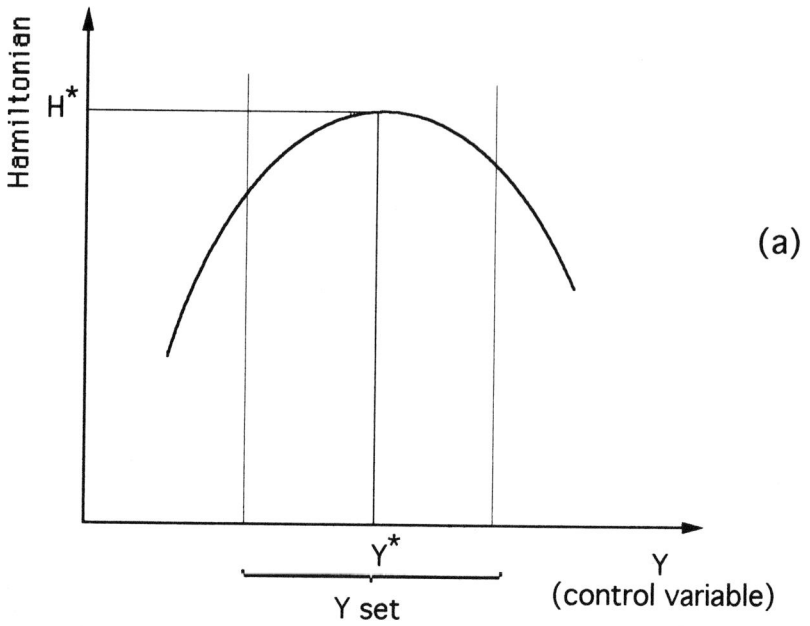

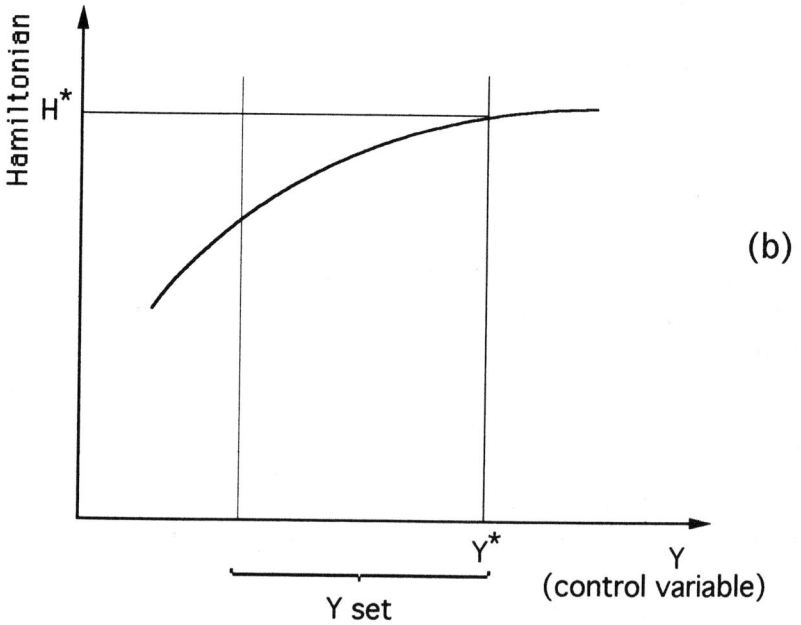

Figure 1.8   Geometric representation for the
              Hamiltonian maximization, interior
              solution (a) and boundary solution (b)

$$\text{max.} \int_0^T U(x(t),y(t))e^{-\delta t} + K(x(t))e^{-\delta t}$$

subject to $\dot{x} = G(x(t),y(t))$ $\qquad x(0)=a$

The Hamiltonian for this problem is:

$$\mathbb{H}(\cdot)=U(\cdot)e^{-\delta t} + \lambda G(\cdot)$$

At this stage the current value Hamiltonian ($\tilde{\mathbb{H}}$) is defined as:

$$\tilde{\mathbb{H}} = \mathbb{H}(\cdot)e^{-\delta t}=U(\cdot) + \mu(t)G(\cdot) \qquad \text{where } \mu(t)= e^{\delta t}\lambda(t)$$

The term "current value" originates from the fact that the Hamiltonian, as defined above, represents a value from the perspective of period t.

The necessary conditions given (eq. 2, page 30) in terms of the Hamiltonian are:

$$\frac{\partial \mathbb{H}(\cdot)}{\partial y(t)} = \frac{\partial U(\cdot)}{\partial y(t)}e^{-\delta t} + \lambda(t)\frac{\partial G(\cdot)}{\partial y(t)} = 0$$

$$\dot{\lambda} = -\frac{\partial \mathbb{H}(\cdot)}{\partial x(t)} = -\frac{\partial U(\cdot)}{\partial x(t)}e^{-\delta t} - \lambda(t)\frac{\partial G(\cdot)}{\partial x(t)}$$

As for the above adjoint equation, according to Dorfman (1969), $-\dot{\lambda}$, it can be understood as the rate of depreciation of capital. Then, the adjoint equation asserts that along the optimal path, this rate of depreciation must equal the sum of marginal flows of capital assets or accumulated dividends. However, this condition hardly seems self-evident.

From the definition of $\mu(t)$ it is implied $\lambda(t) = \mu(t)e^{-\delta t}$ and

$$\dot{\lambda} = -\delta\mu(t)e^{-\delta t} + \dot{\mu}e^{-\delta t}.$$

The above necessary conditions can be rewritten in terms of $\mu(t)$ and $\dot{\mu}$ so that :

$$\frac{\partial U(\cdot)}{\partial y(t)} + \mu(t)\frac{\partial G(\cdot)}{\partial y(t)} = 0 \quad \text{and} \quad \dot{\mu} = -\frac{\partial U(\cdot)}{\partial x(t)} + \mu(t)\left(\delta - \frac{\partial G(\cdot)}{\partial x(t)}\right)$$

The multiplier $\mu(t)$ can be thought of as current-value shadow price and $\lambda(t)$ as the present value of the fishery assets. In other words, $\mu(t)$ is the value of an additional unit of $x(t)$ at instant t, while $\lambda(t)$ is the value that is imputed by an incremental unit in $x(t)$ from the perspective of t=0.

The complete set of first order conditions can be expressed in terms of the current-value Hamiltonian as:

$$\frac{\partial \tilde{H}(\cdot)}{\partial y(t)}=0, \quad \dot{\mu} - \delta\mu(t)=\frac{\partial \tilde{H}(\cdot)}{\partial x(t)}, \quad \dot{x} =\frac{\partial \tilde{H}(\cdot)}{\partial \mu(t)}, \quad \mu(T)=K'(\cdot), \quad x(0)=a$$

Optimal equilibrium and dynamic approach for a fisheries model

The dynamic fisheries optimization problem, of the maximization of the present value of net social benefits:

$$\max. \int_0^T (py(t)-cE)e^{-\delta t}dt$$

$$\text{subject to } \dot{x} = F(x(t)) - y(t) \qquad x(0)=x_0$$

introduced on page 21 is analysed. The control variable $y(t)$ is assumed subject to the constraints $0 \le y(t) \le y_{max}$ where y_{max} may in general be a given function $y_{max}=y_{max}[t,x(t),E]$ and may be viewed as being determined by the fishing industry's capacity to harvest at any point in time.

The current value Hamiltonian for this problem is:

$$\tilde{H}(x(t), y(t), \mu(t))= (py(t) - cE) + \mu[F(x(t))-y(t)]$$

Considering the production function $Y=qXE$ introduced on page 16 and substituting effort, the current value Hamiltonian is given by:

$$\tilde{H}(x(t), y(t), \mu(t))= \left[p - \frac{c}{qx(t)} - \mu(t)\right]y(t) + \mu(t)F(x(t)) =$$

$$=\sigma(t)y(t) + \mu(t)F(x(t))$$

where $\sigma(t) = p - \left(\mu(t) + \frac{c}{qx(t)}\right)$ is called the switching function.

Given that the objective function is linear in the control variable $y(t)$ the problem is a linear optimal control problem that seeks to determine the optimal control y(t)=y*(t), t≥0 and the corresponding optimal population x(t)=x*(t), t≥0.

The necessary conditions of the Maximum Principle y(t) must maximize the $\tilde{H}(\cdot)$ over the allowable range, $0 \leq y(t) \leq y$

This implies that: $y(t) = \begin{cases} 0 & \text{if} \quad \mu(t) > p - \dfrac{c}{qx(t)} \\ y_{max} & \text{if} \quad \mu(t) < p - \dfrac{c}{qx(t)} \end{cases}$

This explains the term switching function for $\sigma(t)$ because it determines whether y(t) should switch from y(t)=0 to y(t)=y$_{max}$. The economic sense of the above condition asserts that the resource should be harvested only when the net revenue per unit harvest (p-c/qx(t)) exceeds the shadow price $\mu(t)$. However, unless $\mu(t)$ can be computed, which is extremely difficult if not impossible (Clark, 1990) by the nature of the industry the above condition could not be a straightforward operational decision rule.

The most important case arises when,

$$\sigma(t)=0, \text{ i.e. } \mu(t)=p - \frac{c}{qx(t)}.$$

This leads to the singular or myopic solution, assuming that $\sigma(t)=0$ holds for t in some interval:

If we differentiate $\dot{\mu} = \frac{c}{qx^2} \dot{x} = \frac{c}{qx^2}(F(x(t))-y(t))$ (EQ. 3)

and apply the adjoint equation, we get:

$$\dot{\mu} = \delta\mu - \tilde{H}_x = \delta\mu - \frac{c}{qx^2}y(t) - \mu F'(x(t)) =$$

$$= -\frac{c}{qx^2}y(t) + \left(p - \frac{c}{qx(t)}\right)(\delta - F'(x(t))) \quad\quad (EQ.\ 4)$$

Equating the right sides of (eq. 3) and (eq. 4) gives:

$$F'(x(t)) + \frac{cF(x(t))}{x(qpx-c)} = \delta$$

In economic terms the above equation would be called the "Golden Rule of Capital Accumulation" (i.e. G.R.C.A.). Clark and Munro (1975) Clark and his colleagues (1986) Pearce and Turner (1990) point that the LHS of the above equilibrium represents the "own rate of interest" of the resource, that is equal to

35

the social rate of discount. The "own rate of interest" consists of two components: the instantaneous marginal product of the fish resource and what Clark and his colleagues (1986) name as the "marginal stock effect" i.e. a measure of the marginal influence of stock abundance upon net revenue or rent flows $U(x(t), y(t))$ and is analogous to the "wealth effect" found in modern capital theory.

The above "golden rule of capital accumulation" (G.R.C.A.) is an implicit equation for the unknown $x(t)$ where in the case of the logistic model (i.e. $F(x(t)) = rx(t)\left(1 - \frac{x(t)}{k}\right)$ see page 16) it reduces to an easily solved quadratic equation with a unique solution $x(t)=x^*(t) > 0$ in most cases of interest (Clark and Munro, 1975). In that case $x^*(t)$ represents the optimal equilibrium stock level. Given that $x^*(t)$ is the unique solution of the G.R.C.A. equation and given an initial population $x(0)$, the optimal harvest policy may then be described as follows:

$$y(t) = \begin{cases} y_{max.} & \text{whenever} & x(t) > x^*(t) \\ y^*(t) = F^*(x(t)) & \text{whenever} & x(t) = x^*(t) \\ 0 & \text{whenever} & x(t) < x^*(t) \end{cases}$$

This solution is an example of the Most Rapid Approach Path (MRAP) (Spence *et al.*, 1975) or a "bang-bang" control (Clark and Munro, 1975). It has an obvious "feedback" form as harvest takes place at its lower or upper bound to return to its optimal level as rapidly as possible (figure 1.9). If the initial stock $x(0) > x^*(t)$ then $y(t) = y_{max.}$ until $x(t) = x^*(t)$ at some $t=t_1$.

MRAP solutions are generally easy to describe and compute, and they can, therefore, be explained to a public with little technical and mathematical knowledge. But of course they only arise under special and possibly unrealistic conditions and circumstances. The elusive assumptions of fixed price and unchanged demand curve do not take account of the frictional effects of market forces, which reduce the value of MRAP as a management decision tool for the resource management. As for the present problem, the essential condition for a MRAP solution, as Spence and his colleagues (1975) postulate, is when the objective function is linear in the control variable $y(t)$, which is satisfied for our present model as can be easily shown. However, such solutions seeking an optimal path $x^*(t)$, whatever the consequences, may appear somewhat unrealistic. For example, if overfishing occurs, to close down a fishery completely in order to bring back the fisheries population to $x^*(t)$ seems to be an extreme act, in particular if the closure is expected to be lengthy.

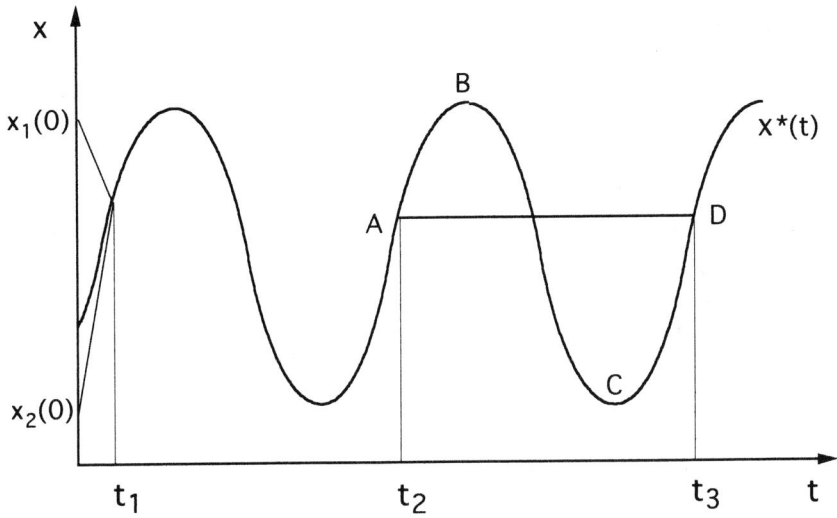

Figure 1.9   Most Rapid Approach solution and
blocked interval for a myopic harvest
policy rule x*(t)

The optimal equilibrium path of the stock, x*(t), is called a singular path and it is derived because the Hamiltonian expression is linear in the control variable y(t). This implies that the singular case σ(t)=0, which holds for some t in some interval, exactly corresponds to the singular solution x(t)=x*(t). Thus, x*(t) determines a unique singular path which is the desired optimal solution. More precisely, the optimal solution of x(t) must be as close to x*(t) as possible and may be approached more or less rapidly depending upon the structure and the parameters of the problem.

The singular optimal biomass x*(t) is sometimes referred to as a myopic decision rule. This terminology, due to Arrow (1964), stems from the observation that the singular path is determined entirely on the current value of parameters and their current time rates of change. No matter how infinite a horizon is adopted in the objective function, the singular solution requires only local-time data. The validity of myopic policies depends upon the possibility of adjusting the stock level x(t) continuously to follow the singular path x*(t), if and when this is possible. Indeed, cases can well arise in which the myopic rule is not feasible because of constraints that prevent x(t) from following the singular path x*(t). In such cases, optimal management usually requires foreknowledge of future parameter shifts, in order to decide when to begin diverging from the myopic path. (Arrow 1964; 1968) have introduced the term "blocked interval" (blocked by the constraint) to refer to any time interval during which the myopic path is not followed and the control variable assumes a constrained value.

For example, let us suppose that there is a blocked interval referring to time (t_2, t_3) (Figure 1.9). During that interval, the constraint $\dot{x} \geq 0$ of our linear control problem prevents the path x(t) from following the singular path x*(t). Then, x(t) cannot follow x*(t) on its downward arc BC but must follow AD. This is because MRAP theory implies that the optimal path must always lie as close as possible to the singular path. This explains that the first switch which occurs at t_2 is "premature" since the myopic path could be followed up to B. In other words, Arrow (1968) states that a digression from a myopic path must begin before this path ceases to be feasible.

When the optimal approach path is determined, the optimal resource investment policy, until x*(t) is achieved, is also determined. Although myopic decision rules are easily treated analytically, they sometimes involve extreme assumptions. Whenever the fish stock, x(t), is below the optimal level x*, the myopic rule requires zero harvesting which could have undesirable consequences for fishermen and processors. Even consumers could also be adversely affected, unless substitute foods are available at competing prices. On the other hand, if the fish stock is above the optimal level $x^*(t)$, there must be an investment at maximum rate until $x^*(t)$ is reached. Once $x^*(t)$ is reached, fishing should proceed at least until a parameter shift occurs. It is unreasonable to assume that prices will remain constant through time or that the cost function will not shift. Demand shifts through time are bound to occur and technological changes are likely to influence costs. In addition, even if blocked intervals were introduced before a myopic path ceased to be feasible, the problem would be still more complicated.

Therefore, MRAP implies straightforward managemental responses that could sometimes be fairly rational in short-run. It is mainly because of the randomness in growth (i.e. the reproduction and mortality) of fish stocks and the dependence of producers in certain regions that myopic and short-run responses are a critical consideration (Kirkley and Strand, 1988). Nevertheless, the linear dependence of both price and costs on harvest rate $y(t)$ does not reflect reality in the long-run and should be dropped, indicating that the myopic path is no longer optimal. Furthermore there is another important, restrictive but implicit assumption throughout the above discussion, that all of the capital used in exploiting the resource is perfectly malleable. This, will be relaxed further below.

Non-Linear Fisheries Models

In order to discuss the theory of myopic decision it was considered useful to develop a linear model where the price, p, of landed fish is constant and independent of the catch rate $y(t)$. This assumption is usually reasonable for small fisheries that deliver their catches to large markets. The small scale fishery is unable to affect market price by raising or lowering the supply of fish. Thus, at the micro level, a small fishery can only decide from time to time whether or not to leave the fishing activity using the myopic decision theory.

However, many fisheries face small local markets where price may be highly sensitive to the current fish supply. For example, this sensitivity is noticeable for "luxury" species such as halibut, which are not substitutable by other species. Furthermore, some local fisheries are so large that their output directly affects prices. For example, the collapse of the Peruvian anchoveta fishery in 1973 affected prices of fish meal and its substitutes (Murphy, 1973; Csirke, 1988).

For a competitive fishing industry, the price received per unit of harvest is no longer constant and varies according to the inverse demand curve, where $D'<0$. Copes (1972) asserts that the introduction of non-linearities into the fisheries model does not necessarily result in the complete dissipation of net economic benefits from the fishery even though an inadequate management is applied. That is because benefits in the form of consumer surplus are great enough to prevent the anticipation of the rising fishing costs.

The social benefit associated with harvest, Y(t), is the area under the demand curve from zero to y(t). That is

$$B(Y(t)) = \int_0^{y(t)} D(y) \, d(y)$$

Suppose that the cost per unit effort is constant or linear in y(t) and the production function is linear in effort, then the socially optimal fishery management policy requires the maximization of the present value of net social benefits from harvesting over the period T and is given by

$$\max \int_0^T [B(y) - cE(t)] \, e^{-\delta t} \, dt$$

subject to the state equation $\dot{x} = F(x) - Y$ and $x(0) = x_0$.

The current value Hamiltonian for this problem, assuming the production function Y=qXE, becomes:

$$\tilde{H}(x(t), Y, \mu(t)) = B(y) - \frac{cY}{qx(t)} + \mu[F(x(t)) - Y]$$

The maximum principle implies that

$$\frac{\partial \tilde{H}(\cdot)}{\partial Y} = 0 \implies \mu = B'(Y) - \frac{c}{qx(t)} = D(Y) - \frac{c}{qx(t)}$$

and differentiating with respect to t, then

$$\dot{\mu} = D'(Y) \, \dot{Y} + \frac{c}{qx^2} [F(x(t)) - Y]$$

The adjoint equation asserts $\dot{\mu} = \delta\mu - \frac{\partial \tilde{H}(\cdot)}{\partial x(t)} = \delta\mu - \frac{cY}{qx(t)^2} - \mu F'(x(t)) =$

$$[\delta - F'(x(t))] \left[D(Y) - \frac{c}{qx(t)} \right] - \frac{cY}{qx(t)^2}$$

Equating and solving for $\dot{Y}$ we obtain

$$\dot{Y} = \frac{1}{D'(Y)} \left[\left[\delta - F'x(t) \right] \left[D(Y) - \frac{c}{qx(t)} \right] - \frac{c}{qx(t)^2} Fx(t) \right] \quad \text{[EQ. 6]}$$

and $\dot{x} = F(x(t)) - Y$.

According to the maximum principle, any optimal trajectory $(x^*(t), y^*(t))$, must satisfy the non-linear autonomous system [eq. 6]. In order to find the optimal equilibrium solution we must set $\dot{x} = \dot{y} = 0$. This gives

$$Y = F(x(t)) \quad \text{and} \quad F'(x(t)) + \frac{cF(x(t))}{x(t)[qx(t)D(F(x(t))) - c]} = \delta \quad \text{[EQ. 7]}$$

In that case we obtain a similar implicit equation to [eq. 5] for the optimal steady state $x(t) = x^*(t)$. Indeed the two equations become identical in the event that $D(Y) = p$.

The optimal approach path to the equilibrium solution $(x^*(t), y^*(t))$ can be obtained by solving the system of [eq. 6]. Once a steady state solution $(x^*(t), y^*(t))$ for the dynamic optimization problem has been identified, it is preferable to determine its local stability, that is the dynamic behaviour of the system near the steady state equilibrium. This requires an analysis of the solution trajectories in a phase-plane map of the system. The behaviour of dynamical systems proposed as models of ecosystems is analysed in detail by May, (1975).

For the present model the x-y plane is the phase-plane, where we could proceed by drawing the isoclines and identifying the directionals in the various isosectors. There are many examples (Miller, 1970) of the use of phase plane diagrams for the analysis of systems of nonlinear differential equations that govern the motion of variables in dynamic economic models. Such equations are at the heart of models of economic growth (Burmeister and Dobell, 1970). In the case of fisheries several authors (Arrow, 1970; Clark, 1990, and more recently Neher, 1990) underline the usefulness of phase-plane diagrams in optimal economic growth model analysis. Clark (1990) analysed data from the pacific Halibut fishery and used a model similar to [eq. 6] to show graphically the dynamics of this particular fishery (figure 1.10). The equilibrium, $(x^*(t), y^*(t))$ is the intersection of two isoclines $\dot{x} = 0$ and $\dot{y} = 0$. From [eq. 7] it can be derived that the x-isocline is the known growth curve from the Schaefer model $Y^* = F(x^*(t))$ and the y-isocline is the line $x(t) = x^*(t)$,

$$\text{where} \quad F'(x^*(t)) = \delta - \frac{cF(x^*(t))}{x^*(t)[qx^*(t)D(F(x^*(t))) - c]}$$

The isoclines divide the orthant into four isosectors. Each isosector has a directional that indicates the movement of a point (x,y) over time. For every

41

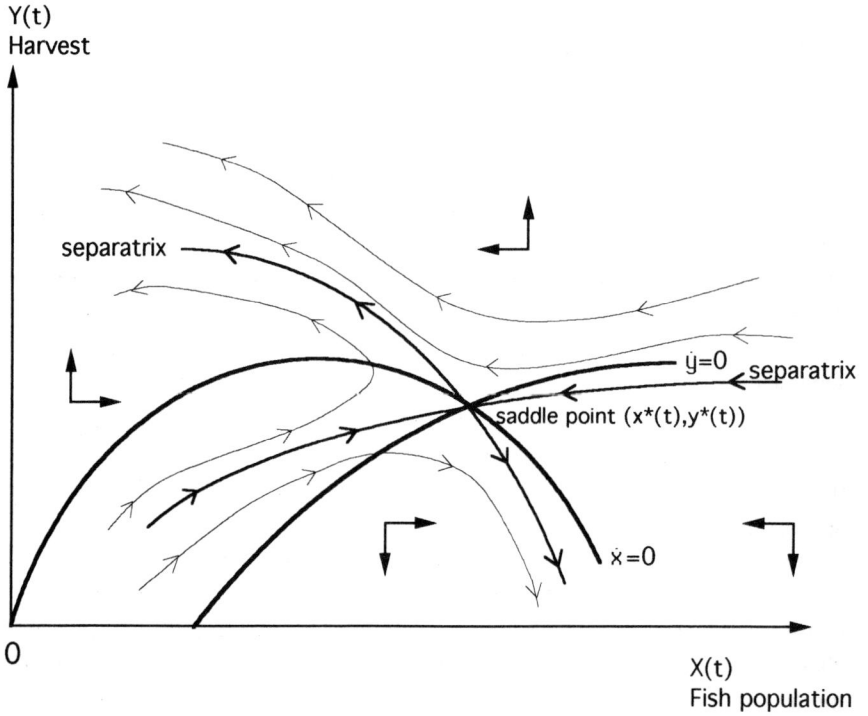

Figure 1.10   Phase-plane diagram for the analysis
of a nonlinear fishery model

initial point (x_0, y_0), [eq. 6] possesses a unique solution $(x(t), y(t))$. Any solution, $(x(t), y(t))$, is called a trajectory or an internal curve of the system. Through each given point (x, y) there passes a unique trajectory. The geometrical analysis of each trajectory is facilitated by considering the isoclines. That is because any trajectory must cross the $\dot{y} = 0$ isocline horizontally and the $\dot{x} = 0$ isocline vertically, except for the point where the two isoclines intersect, that is the equilibrium point of the system. The particular trajectories that pass through the equilibrium point are called separatrices. Taken together, the two separatrices define the optimal solution trajectories for our infinite horizon problem and determine the local stability of the equilibrium point, which depends on the convergence or divergence of the separatrices toward this point. In general the separatrices must be computed numerically. However, it must be noted here that even with ordinary algebraic equations a dynamic system is often not solvable. Although numerical techniques do exist (eg. internal bisection, Newton's method, eigenvalues), in many cases it is even impossible to obtain a geometrical description of the trajectories. For the present model, the separatrices associated with the equilibrium point either converge towards $(x^*(t), y^*(t))$ as $t \to \infty$ or diverge from $(x^*(t), y^*(t))$ as t increases from $-\infty$. Such a point in dynamical system terms is classified as a saddle point equilibrium for the system (figure 1.10). If the separatrices can be calculated, the optimal approach path follows the separatrices converging to $(x^*(t), y^*(t))$. In that case, the separatrices define an optimal feedback control for our problem; that is Y is specified for each stock level $x(t)$ by these curves. Thus, the optimal approach equilibrium is no longer a MRAP.

The significance of using phase-plane diagrams for a non-linear fishery model, as described above, is shown by the fact that the optimal approach path is a more gradual approach, engendered by market reactions to the harvest rate, than is the myopic decision path. Better, in the linear model where the revenues from fishing are assumed to be directly proportional to harvest, the MRAP consisted only of adjusting $x(t)$ as rapidly as possible to the equilibrium stock level $x^*(t)$. Problems, such as "flooding the market" and driving down the price of fish by harvesting at a high rate, or, on the other hand, low harvest rates leading to increased prices were not assumed. In contrast to myopic decision paths, the above non-linear model introduces such phenomena.

Overall, the methodology of obtaining an optimal approach path for a fishery model such as the above is to choose some trajectory $(x(t), y(t))$ passing through a given initial point (x_0, y_0) that converges to $(x^*(t), y^*(t))$ as closely as possible. The ideal trajectory would be the separatrix that steadily converges to $(x^*(t), y^*(t))$, because all the other trajectories first approach the saddle point (x*(t),y*(t)) but then diverge from it. An important criterion for a trajectory to become the optimal approach path is time. Depending on the length of time, T, that is taken to traverse a given trajectory we can choose a trajectory passing closer to the equilibrium point. That is because the velocity of points on trajectories near (x*(t),y*(t)) is smaller than the velocity of points on other trajectories. Therefore, if the available time, T_1, to optimize our fisheries model is relatively small, then the required optimal trajectory remains far from (x*(t),y*(t)). Conversely, if T_2, is ample enough then we have the time to choose the trajectory that lies near the separatrices and the point (x*(t),y*(t)). In this case the optimal approach path spends most of its time in the vicinity of (x*(t),y*(t)) (Samuelson, 1965) (figure 1.10).

Non-malleability of capital and optimal investment strategies in fisheries management

In all fisheries models previously described, fishing costs have been assumed to be directly proportional to effort. Since effort is variable, this means that only variable costs have been taken into consideration. Clark (1985) estimated that roughly 25% of the fishing costs per vessel for a small scallop fishery in New Zealand are fixed, regardless of the amount of fishing actually undertaken. Actually, many fisheries are much more capital intensive than this example. In general, fisheries require vessels and gear, port facilities and processing plants. Such investments are non-malleable in the sense that there are constraints upon the disinvestment of capital assets. These investment costs cannot be recovered later if equipment is no longer needed (Arrow, 1968).

In the previously described dynamic models, the implicit assumption that capital was perfectly malleable offered remarkable analytical advantages. The analysis was facilitated, as it was concerned with only one investment problem, that is the investment in the resource itself. This was reflected in the application of optimal control theory to fisheries and as a consequence only one relatively tractable state variable problem was addressed. With capital assumed to be non-malleable we are confronted with a considerably more complex problem, since it demands a minimum of two state variables.

An example of the influence of non-malleable capital upon optimal fisheries management, based on Clark and his colleagues (1979) model, is now discussed (Eqs. 8-11).

The Schaefer population dynamics model for a fishery resource is given in eq. 8.

$$\dot{x} = F(x(t)) - qE(t)x(t) \qquad (EQ. 8)$$

44

In this case it is more convenient to use E(t) rather than Y(t) (see page 18) as the control variable relevant to x(t).

In (eq. 9) E(t) is subjected to constraints where E_{max} is the maximum effort that can be exerted and is taken to be proportinal to the size of fishing fleet K(t), or else the amount of capital invested in the fishery at time t.

$$0 \leq E(t) \leq E_{max}=K(t) \qquad (EQ.\ 9)$$

If I(t) is the rate of gross investment in the fleet and $\gamma(t)$ the rate of depreciation then the change of the capital investment, K, in time becomes,

$$\dot{K} = I(t) - \gamma K(t) \qquad (EQ.\ 10).$$

The non-malleability assumption is incorporated in eq. 11, as there is a non-negativity constraint imposed on the investment rate I(t), as well as a non obvious upper bound on investment.

$$I \geq 0 \qquad (EQ.\ 11)$$

Therefore, I(t) is allowed to be arbitrarily large or even infinite. This amounts to the assumption that there is no resale market whatsoever for excess capacity.

The fisheries manager wishes to determine an investment strategy I(t) and an effort strategy E(t). The amount of fishing effort that can be exerted is proportional to the size of the fishing fleet, or in other words to the capital K(t) (eq. 9). The maximization of the present value of net revenues requires

$$\max_{E(t)\ I(t)} \int_0^\infty e^{-\delta t}[pY-cE-c_f I]dt$$

subjected to conditions described by eq. 8-11 for t≥0, where c_f denotes the unit cost of capital invested.

As we have previously seen, since
$Y = qxE$, the maximand becomes

$$\max_{E(t)\ I(t)} \int_0^\infty e^{-\delta t}[(pqX-c)E-c_f I]dt.$$

The above control problem is a two state variable [x(t), K(t)] and two control variable [E(t), I(t)] problem. The feedback control for the optimal harvest and investment policy is illustrated in (figure 1.11), a state space diagram x-K divided into three regions R_1, R_2, R_3 by two curves σ_1, σ_2. Figure 1.11 specifies the entire sequence of investment and harvesting strategies and gives the optimal values of I(t) and E(t) for the points in these three regions. The

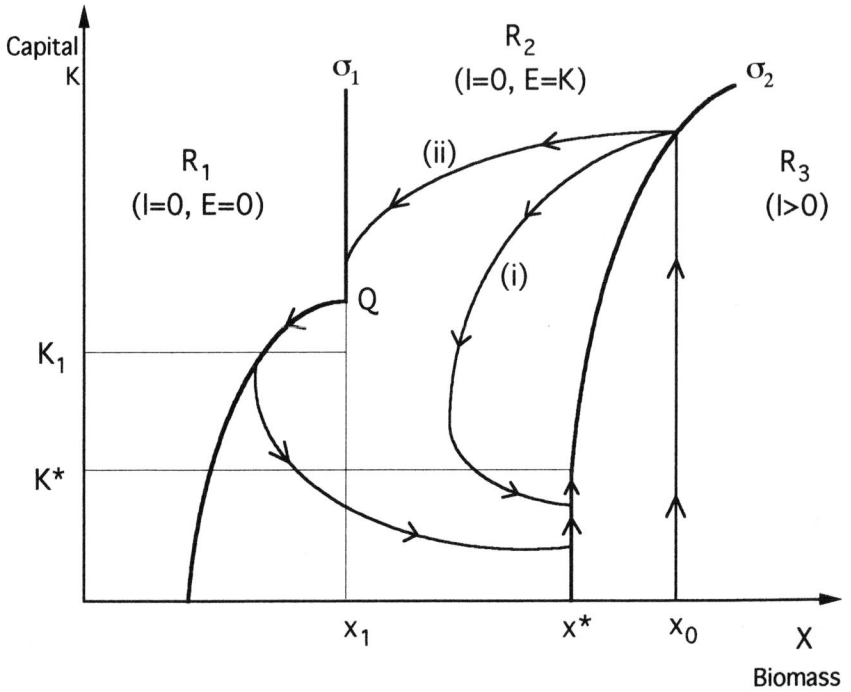

Figure 1.11   Feedback control: diagrammatic
               representation of optimal harvest and
               investment policies

curves σ_1, σ_2 are called switching curves (see below) and can be computed numerically.

A fisheries manager can determine the optimal investment/effort policy starting from the current values of $x(t)$ and $K(t)$ and choosing his policy according to the rules:

1) If $(x(t), K(t))$ is in the R_1 area, then no investment must be applied and the fishing activities must be discontinued ($E(t) = 0$). Thus, σ_1 specifies the biomass levels below which E should switch to zero.

2) If $(x(t),K(t))$ is in the R_2 area, then no investment is required and fishing must be applied at full capacity ($E(t) = K(t)$) for an optimal policy.

3) If $(x(t),K(t))$ is in the R_3 area, then positive gross investment in the fleet can occur ($I(t)>0$) and K can be increased up to a value determined by σ_2 curve. Thus, the switching curve σ_2 specifies the appropriate level of gross investment in the fleet.

For example, let us suppose that the initial stock level is $x_0>x^*$ then initial investment has to be made. Focusing on the dynamic behaviour of the system, all trajectories in the R_2 region move downwards with the passage of time because of the depreciation of capital. Then, employing full capacity $E(t) \equiv K(t)$ in the region R_2, a MRAP is used and the target level of fish stock becomes x_1. There are in fact two cases: (see figure 1.11) i) if K_0 is small, $x(t)$ may never reach x_1, and ii) if K_0 is large, $x(t)$ will reach x_1 and will stay there temporarily. When x_1 is reached, capital is "abundant" and optimal policy calls for harvesting x_1 on a sustained yield basis. Indeed, a temporary sustained yield harvesting occurs along the σ_1 curve above the point Q (Clark *et al.*, 1979). As has been mentioned the policy is temporary as the capital "abundance" is temporary. When the switching curve σ_1 meets the line $x(t)=x_1$ at a point Q above $K=K_1$ the depreciation rate is positive and it is non-optimal to aquire new vessels as long as $x(t)<x^*$, thus, K must fall below K_1. Therefore, a switch in the harvesting policy occurs leading to a path targeting to $x(t)=x^*$, even though this switch may cause $x(t)$ to fall temporarily below the optimum x_1 (see figure 1.11). When $x(t)$ lies in the region R_2 the optimal policy utilizes full capacity $E(t)=K(t)$ but no new investment is undertaken. Finally, the stock reaches $x(t)=x^*$, when additional investment up to K^* is made and the long run equilibrium is reached at $x(t)=x^*$, $K=K^*$, $E=K^*$ and $I=\gamma K^*$ with the implication that investment covers depreciation. An application of the above model has been fitted to data from the Antarctic baleen whale fishery by Clark and Lamberson (1982).

47

Clark and his colleagues (1979) assert that the long run equilibrium at (x^*,K^*) coincides with the case of perfectly malleable capital (see previous models above). However, as the fisheries managers attempt to enforce a programme for resource conservation, they must distinguish between the two restoration policies which are different. In the "non-malleable case" as opposed to the "malleable case", the optimal policy requires a far more gradual adjustment approach rather than the "bang-bang" approach for the "malleable case" described previously. Draconian policies, such as shutting the fishery down entirely, in implementing a MRAP for the "malleable" case, take no consideration of the fact that the fishing vessels may have no other viable alternative uses. As for the "non-malleable" case an optimal management programme calls only initially for a harvest moratorium, and as stock level reaches x_1, open access economic forces, taking into consideration the non-maleability of capital and the depreciation rate, will lead to $x(t)=x^*$ and finally at the long run equilibrium (x^*, K^*). Then, in order to remain at (x^*, K^*) there must be a continual replacement of the depreciating capital at the rate $I=\gamma K^*$.

A problem of practical importance for the management of fisheries when stocks have become depleted is that the simplistic ban on fishing activities may be no longer applicable The above analysis shows that the concept of depletion depends, in a rather complicated way, on the economic aspects of the fishery. Relative to the long-run optimal equilibrium, x^*, a fish stock may be considered depleted although the fish stock lies somewhere between x_1 and x^*. Then, as the model demonstrates the fishery should remain open but any addition to capacity must be rigorously prevented.

In the case of excessive fishing capacity the above model can be altered to allow a "buyback" programme to remove vessels from the fishery. In order to create a positive opportunity cost for the excess capacity to have a positive value in alternative uses we can remove the constraint $I(t) \geq 0$ assuming that unwanted capacity can be sold off at a price c_s where $c_s < c_f$.

Thus, the objective becomes,

$$\max_{E(t)\ I(t)} \int_0^\infty e^{-\delta t}\left[(pqX-c)E-\Phi I\right]dt$$

$$\text{with } \Phi = \begin{cases} c_f & \text{if } I > 0 \\ c_s & \text{if } I < 0 \end{cases}$$

The solution can be illustrated in the previous figure where in the region R_1 disinvestment $(I<0)$ must occur whenever $(x(t), K(t))$ falls into this region.

In some cases there are variable costs which are mainly proportional to K rather than to fishing effort E. For example, vessels may have to travel considerable distances to the fishing ground or rush from one fishing ground to another, following the opening of the fishery. Fraser (1980) refers to the British Columbia herring roe fishery, where vessels have sometimes been transported by helicopters from one fishing area to another.

As a conclusion to the above models for non-malleabe capital in the fishing industry we can assert that although investment problems are important, they are by nature complicated. The analysis with dynamic optimization techniques to problems of this type becomes additionally complicated when more than one state variable and one control variable are involved. A complete solution to a multidimensional control problem is usually a major research task and normally requires expert mathematical advice. These complications reinstate the value of the study of simplified models.

Conclusions

According to article 39(1) of The Rome Treaty the European Community fisheries regulators should aim for efficiency, equity and price stability. This chapter has critically explored the contribution of fisheries economics in establishing criteria for the implementation of resource management.

The salient conclusion of this chapter is that mainly as a consequence of the lack of ownership of a factor of production (i.e. the fishing ground) in an open-access fishery the economic rent tends to be dissipated. This conclusion is independent of time preference. When account is taken of time preference a further difficulty may arise if the rate of discount exceeds the natural growth rate of the fish stock. For then the stock may be fished to extinction (Clark, 1985). This is further emphasised by the most rapid approach of optimal harvest policies under open access where profit maximizing requires the harvesting of the population as fast as possible. Conversely, when population is overexploited the optimal harvest policy indicates the closure of the fishery until the stock recovers. This strategy is likely to be employed in extreme circumstances when there is a severe decline of stocks. For example the complete closure for the herring fisheries was enforced in the North Sea and West Coast fishing grounds during the years 1977 to 1983 and currently for the Northern cod fishery off Newfoundland.

The economic models have developed from the basic static Gordon model to the analysis of dynamic non-linear fisheries models, which employ optimization techniques. As models become more complicated they overcome the problem of need to make too many assumptions and they tend to represent reality more accurately. However, since they employ many parameters and sophisticated mathematical techniques they do not guarantee an efficient means of managerial strategy formation. In that respect it could be that simpler models are more effective in tackling practical fishery problems. Dynamic analysis enables economists to explore the adjustment phases of measures employed by fisheries managers. Despite the various disputes over the relative importance of complex mathematical optimization models it should be recognised that they offer an approach, which allows for unique perceptions, and predictions which are distinguished from biological or other approaches.

In formulating the CFP the role of economics has been neglected. Recently, at the end of 1992, the Community replaced the Scientific and Technical Committee for Fisheries (STCF) with the Scientific, Technical and Economic Committee for Fisheries (STECF). This Committee is to be consulted on the biological, technical and economic factors concerning the fishery resource. Its role and the influence of economic advice on the formation of the CFP remain to be seen in future.

It is indisputable that in an area, where a living resource is concerned, the biological characteristics of fish populations have to be taken into consideration when managing the resource. This has led to the development of various population dynamics models, where biologists are called to determine the population characteristics of the stocks concerned and predict future trends. The development of population dynamics, the underlying assumptions and limitations of specific methods and their application to E.C. fisheries management will be discussed in the next chapter.

2 Population dynamics and contemporary fisheries management

Fluctuations in the abundance of fish stocks have been scientifically studied since the middle of 19th century. Data collections have been established to provide an approach to the potential limits of the resources with a view to their use in future conservation and management. Total recorded landings have varied significantly since the earliest records of the mid 19th century with respect to the geographically artificial boundaries of major fishing areas (Anon., 1990). The growing concern about falling catch rates from fisheries which were subject to industrial scale fishing impelled the establishment of many major research institutes in Europe and North America, as well as the creation of the International Council for the Exploration of the Sea (ICES) in 1902 in Copenhagen.

Fish stock fluctuations have been ascribed to various factors such as the effects of environmental parameter variation, the influence of pollution, eutrophication or habitat modification, interactions among marine species in an ecosystem and routine fluctuations in recruitment. In addition to the above, fish stock variability is reflected in fish catches and is sometimes due to the relative wealth of society and investment strategies with respect to fishing vessels, instrumentation and processing equipment (Smith, 1988). Of equally significant importance are decisions of governmental or international organizations (E.C., F.A.O.). Consequently, many hypotheses have been constructed in order to explain the decrease or increase in the abundance of particular fish stocks. However, the explanations given do not remain stable, are tentative and not predictive, a fact that suggests that stock abundance variability is far more complex than a simple cause-effect argument.

Spectacular collapses of fish population levels attracted extensive attention in the 1960s and the early 1970s, and attempts at finding the causes of stock decline were made by scientists (Glantz and Thompson, 1981). The most

51

dramatic collapse was that of the Peruvian anchovetta, but there were significant declines in the clupeoid stocks in Pacific sardine and in North Sea herring (Murphy, 1973). Unexpected increases or explosions in fish population levels followed previous collapses in some cases. For example, Troadec and his colleagues (1980) and Gulland and Garcia (1984) reported a so-called explosion of the trigger fish, Balistes, off West Africa in 1982. In 1978-79 the catch of trigger fish off Guinea-Bissau had been zero. Nevertheless, by 1982 64% of total landings were trigger fish and the population seemed to have spread along the African coastline. At that time, a parallel decline of heavily fished sparids was observed, giving ground to the theory that a "replacement phenomenon" of sparids by trigger fish occurred (Daan, 1980). Caddy and Gulland (1983) surveyed the historical data of a number of exploited stocks from throughout the world and they classified the populations according to their historical patterns into the following categories (figure 2.1):

I   *Steady state.* Stocks where a more or less steady yield seems to be sustainable over reasonably long periods of time.
II   *Cyclical.* Stocks that show strong cyclical yields, with periods of high catch regularly followed by periods of low catch.
III *Irregular.* Stocks that show irregular periods of high abundance, without the consistency in alternation between abundance and scarcity shown in cyclical stocks.
IV   *Spasmodic.* Stocks that have produced major yields and then collapsed without any major recovery.

The above classifications can only give a general idea on how exploited populations behave since many other alternative categories could be invented. The major point that arises from this scheme is that most stocks do not appear capable of producing a steady sustainable yield. The importance of this observation seems to be neglected by the main body of the methodology of stock assessment and of fisheries management theory, where there has been a preoccupation with equilibrium calculations. Nevertheless, there is no evidence to support any stock assessment and management technique on how cyclical, irregular or spasmodic stocks can be best managed.

Overall, although early scientific attention to population fluctuations was concentrated on steep declines, there is ample evidence to the effect that collapses may not be permanent but rather reversible and, furthermore, populations are as likely to explode as to collapse (Garrod, 1988). In the recent review of the state of world fishery resources by the F.A.O. (Anon., 1990), there is evidence that fishery resources around the world are close to their maximum catch limits. Scientific monitoring and scientific methods of

I Steady or predictable fisheries

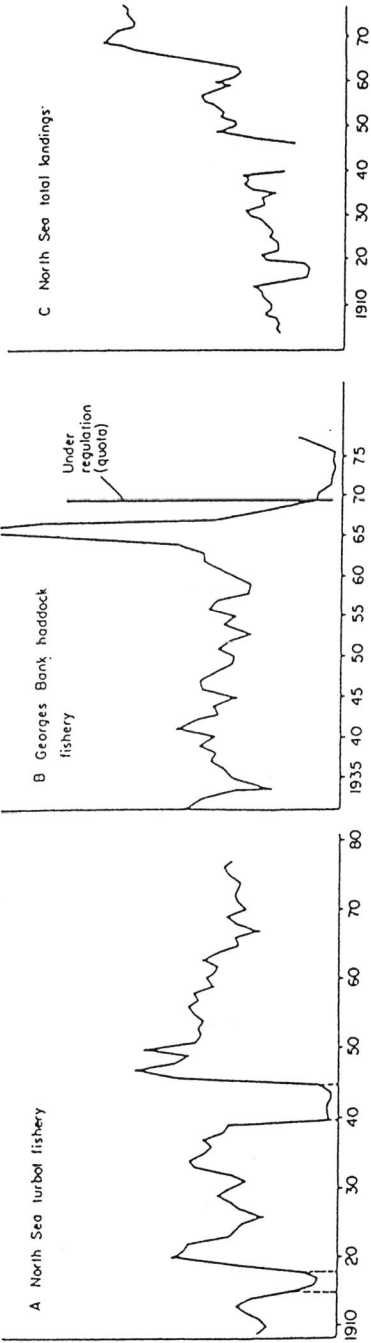

A North Sea turbot fishery

B Georges Bank haddock fishery

C North Sea total landings

II Cyclical fisheries

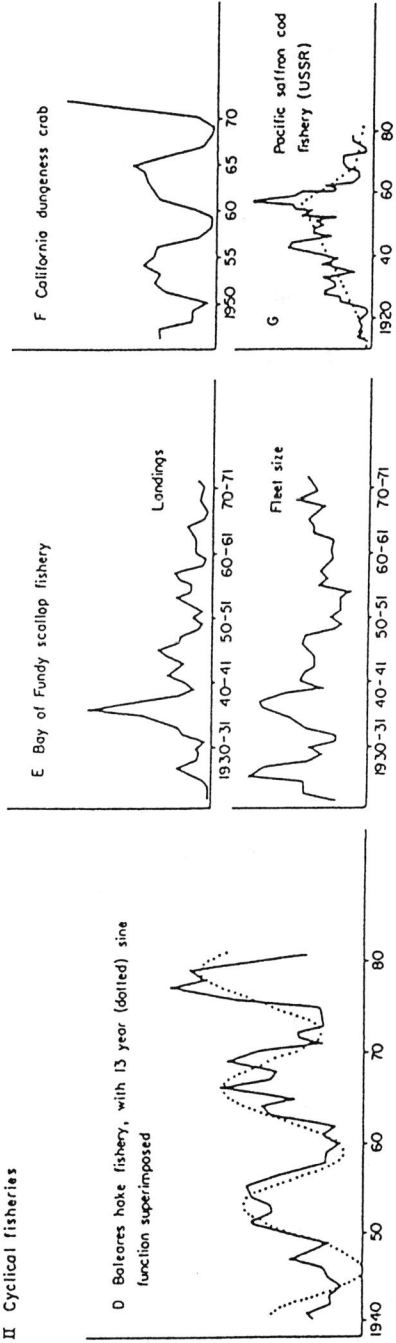

D Baleares hake fishery, with 13 year (dotted) sine function superimposed

E Bay of Fundy scallop fishery

Landings

Fleet size

F California dungeness crab

G Pacific salmon cod fishery (USSR)

53

III Irregular stocks

H Norwegian fishery for juvenile herring

Ring netting

1 year fish

2 year fish

1930 35 40 45 50 55 60 65 70

I Magdalen Islands lobster fishery

1912 21 31 41 51 61 71 81

J Georges Bank scallop
(A–D: period of new fishing investment)

A
II

B
II

C
II

D
II

1930 38 46 54 62 70 78

IV Spasmodic stocks

K California sardine and anchovy fishery

Anchovy

Sardine

1920 30 40 50 60 70

L NW Pacific sardine fisheries

Far East total

Japan

1905 15 25 35 45 55 65 75

M Gulf of Maine shrimp fishery

Landings

Temp.

Effort

1940 50 60 70

N ICNAF area pelagic fisheries

Total ICNAF area

Capelin
Herring
Mackerel

1950 55 60 65 70 75 80

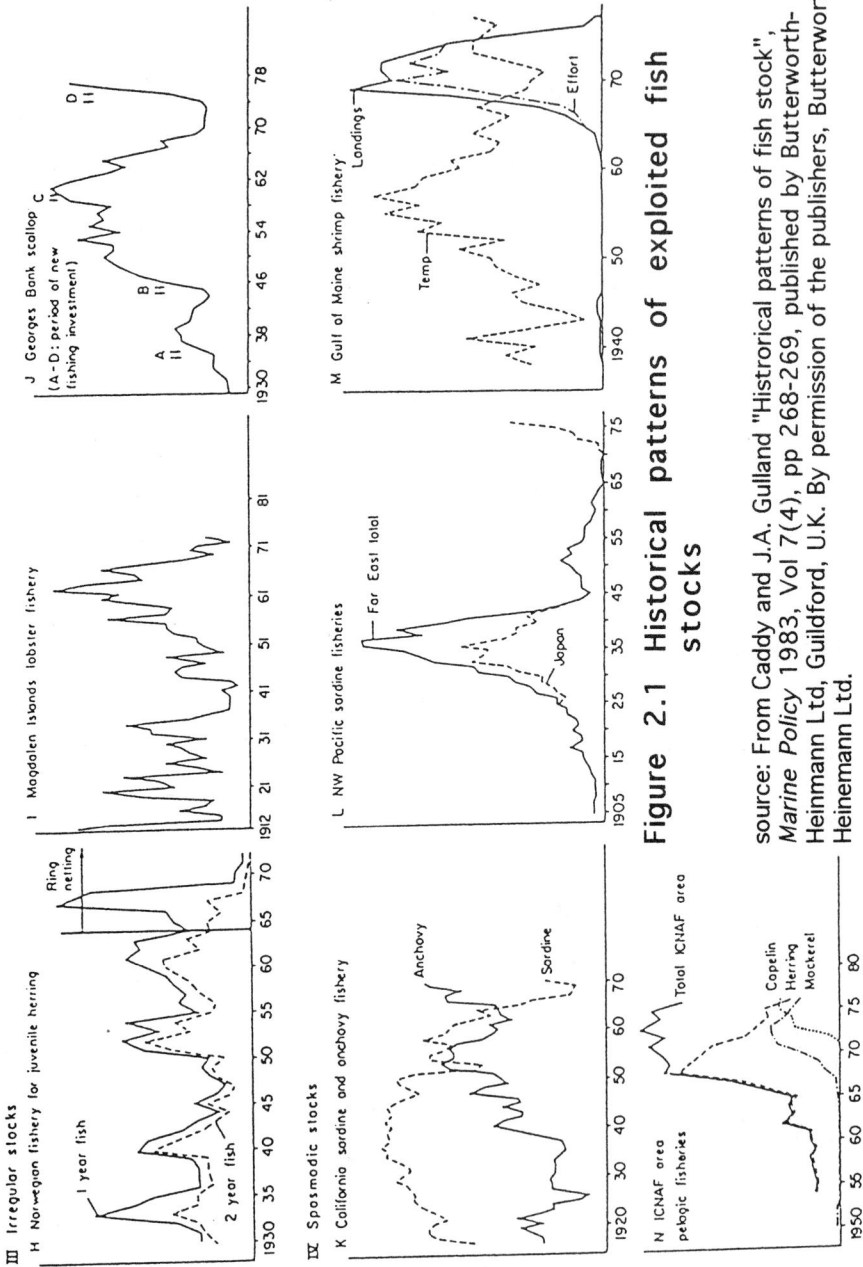

Figure 2.1 Historical patterns of exploited fish stocks

source: From Caddy and J.A. Gulland "Historical patterns of fish stock", *Marine Policy* 1983, Vol 7(4), pp 268-269, published by Butterworth-Heinmann Ltd, Guildford, U.K. By permission of the publishers, Butterworth-Heinemann Ltd.

assessing fish population dynamics have shown that although many fish populations reveal biological degradation and consequently economic waste arising from fishing, landings appear to be more the result of climatic fluctuations on stock sizes. Despite the high quality of the research base for certain fish stocks in certain areas such as for North Sea mackerel (Kennedy, 1989) and North Sea haddock (Horwood, 1991), the relative failure of international management to establish sustainable fisheries is in many cases due to socioeconomic reasons; that is excessive catching capacity and the uncontrolled transfers of fishing effort have been as important as the unpredictability due to biological and ecological phenomena.

Whenever the objective is capture optimization for a certain species, this implies the extraction of resources from the marine ecosystem without affecting it. There is always antagonism between the intensity of fishing allowed and the abundance of resources. In addition to this, problems arise for non-targeted species, which sometimes are already in danger of depletion and are taken as by-catch. Nevertheless, in practice, scientific knowledge does not always seem to be translated into effective management decisions.

Stock recruitment models

Fish population collapses, explosions and periodicities are reflected in time series data for recruitment. Recruitment variability has been the centre of fisheries research, as numerous environmental and biological factors can affect recruitment levels. In addition, observations of the magnitude and behaviour of fish variability certify that the most important component in variability is recruitment (Gulland, 1982; Sissenwine, 1984). However, the problem frequently arises that there are insufficient time series data on recruitment to derive predictions of the occurrence of unusual events such as collapses (Shepherd, 1988).

There has been an attempt to classify different stock-recruitment models into categories as discrete scientific bases for fishery management decisions. Such methodology obscures the pragmatic polyparametric population dynamics of fish stocks, and following one or another model may be detrimental in management decision reasoning. As each model highlights different biotic or abiotic factors, (such as age-specific growth, mortality rates, recruitment and spawner-recruitment interactions) it understates the influence of other variables. On the other hand, general models that treat fish population dynamics as varying aggregations of many components of production (Schnute and Richards, 1990) may obscure the true nature of fish population dynamics due

55

to the "curse of dimensionality". Furthermore, these models may be addressed to certain specialised scientists and, therefore, they may not have any validity for generalised fisheries managers.

J. A. Gulland, one of the leading fisheries scientists (Chapman, 1990) has often opposed sophisticated analyses and complicated calculations in fisheries models on the grounds that they obfuscate rather than clarify the natural processes of growth and reproduction in fish. Such sophisticated analyses often bring a management response of bewilderment in the face of chaotic uncertainties and give grounds to ruthless politicians to diminish fruitful results of fisheries science.

A fundamental stock recruitment model

A general schematic model of fish population dynamics is shown below

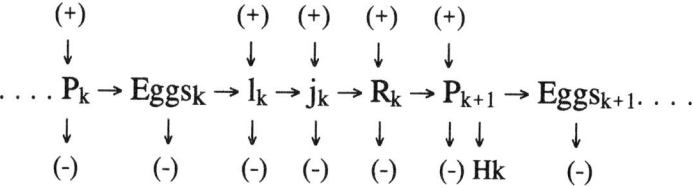

$$\ldots\ldots P_k \to Eggs_k \to l_k \to j_k \to R_k \to P_{k+1} \to Eggs_{k+1}\ldots\ldots$$

where P_k is the parent stock of the kth generation that spawn $Eggs_k$, which hatch. The larvae, l_k, become the juveniles, j_k, and ultimately provide a certain number of recruits, R_k, that may be harvested, Hk, and the remainder give rise to the parent stock P_{k+1} of the k+1 generation. The input (+) to each cross-current flow is ingestion and the output (-) is associated with predation, natural mortality and metabolic losses.

The complexity of fish dynamics is evident even in the above generic model. Above all, natural biological populations are subject to complex dynamic processes that are generally subject to strong seasonal or periodic influences. In contrast to continuous time models, various lagged effects such as the response of fish populations to natural mortality and fishing mortality can be described better in discrete time models.

Age-structured models usually suppose that $t_{k+1}-t_k$ represents a life cycle of a fish population. Usually, a life cycle is supposed to be a calendar year and the group of fish born in a single year is called an age class or a cohort. A discrete time model, according to the above general schematic production model of fish population dynamics is, $R_{k+1}=f(S_k)=f(R_k-H_k)$, where R_{k+1} is interpreted as the recruitment of the population in the k+1 cohort and $S_k=R_k-H_k$ represents what fish biologists refer to as escapement.

56

The above equation is the Ricker stock-recruitment model, developed by Ricker (1954) for use in salmon management. The model presupposes that there is not any overlapping of successive generations. For example, it is assumed that the parent population P_k dies during spawning and is replaced finally by the subsequent cohort of recruits. This type of modelling is used for the management of certain species of Pacific salmon. Nevertheless, the life cycle of Pacific salmon is not one year but varies from two to five years (Larkin, 1988). Despite that, Ricker's model is applied to fish stocks with distinct cohorts; it can be also used for species with overlapping generations but which reach sexual maturity at one year of age. In this case R_k denotes the adult breeding stock. In addition to this, cohorts may be separated in space, as in migrating species. For example, tropical penaeid prawns move offshore as they grow (Garcia, 1988), and, in the Atlantic surf clam fishery, clam beds seem to be dominated by a single cohort (Lovejoy, 1986).

A biological and ecological approach to the stock-recruitment relationship

The variability of stock populations, which, in turn, affects the spawning biomass that determines the magnitude of recruitment, has been previously discussed in such a way that apparently a stock-recruitment relationship must be of indisputable value in population control for future management strategies. Despite the apparent importance of the stock-recruitment relationship, Garrod (1982) has showed that the bulk of empirical data for many fish stocks (e.g. Norwegian and North Sea herring, Acto-Norwegian cod, Arctic and North Sea haddock, North Sea plaice and sole) does not reflect a striking relationship between the magnitude of recruitment and the spawning stock. An explanation of this paradox requires more investigation of the complexities of stock-recruitment theory, and this can be gained only by considering the biological and ecological background to such a mechanism.

Species with short pelagic life-times usually produce a few larvae from eggs rich in yolk (lecithotrophic) and recruitment does not vary greatly from year to year. This is defined by Clarke (1979) as K-selection for maximum competitive advantage in a space or other resource-limited environment. This reproductive strategy is characteristic of cold water species. On the other hand, Vance (1973) has suggested that strategies of r-selection for the maximum rate of population increase are desirable, when food availability for pelagic larvae is unpredictable. Thus, fish larvae with a long planktonic phase are usually produced in large numbers and survival depends on a variety of environmental factors and not on the numbers of the parent stock.

A useful representation of the stock-recruitment curve relating to the above phenomena is suggested by Anon. (1980), figure 2.2. The broad region II in the figure depicts a situation where changes in the recruitment numbers are independent of the current stock size. Such a relationship between stock and recruitment is said to be density independent. A major role in determining density independent models is given to environmental factors. The narrower region I asserts that a density dependence between stock and recruitment exists such that, at low stock densities, the number of recruits will be proportional to the size of the parent stock.

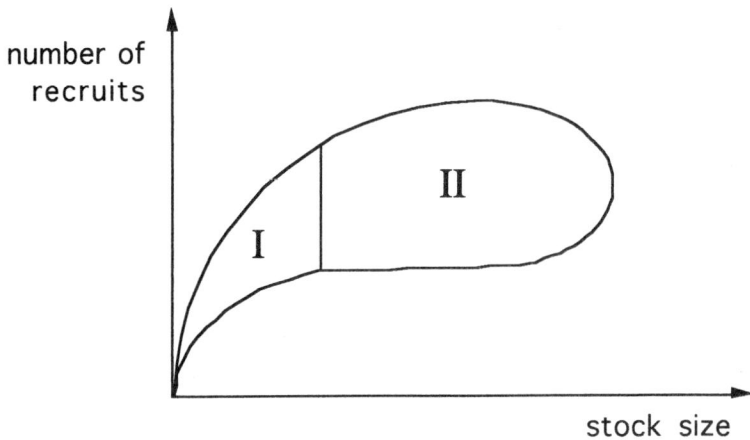

Figure 2.2 General relationship between stock and recruitment. Density dependent region (I) and density independent region (II)

In region II it is assumed that fish are primarily r-strategists and lay large numbers of eggs from which a few adults survive. For example, Cushing (1975) estimates that the mortality for plaice is as high as 80% per month for larvae up to the time of first feeding at 4 months, 10% per month during the first winter when fish are one year old and 10% per year during the life of adult fish. In addition, Anon. (1965) has estimated that from several million eggs laid by one female cod the early life mortality must be about 99.999%. The importance of such a high mortality rate, particularly at the beginning of animal life (planktonic stage), lies not in the mortality percentage of the many but rather in the variable survival of the few. In spite of the high mortality rates of larvae, large numbers of larvae survive.

Although examples of density-dependent growth are rare, Deriso (1985) has attributed the increase in weight of eight-year-old Pacific halibut by 2.5 times to the simultaneous reduction in their numbers by a factor of four. In addition, Cook and Armstrong (1986) have observed density-dependent growth in North Sea haddock and whiting. In this case, a high total stock biomass caused low recruitment. Alternatively, Horwood and his colleagues (1986) have shown that plaice of the southern North Sea stock exhibit a variation in fecundity at a relatively stable stock level. Contrary to the expectation from compensatory density-dependent growth, they found that during 1951-5 and 1975-80 fecundity increased over time by 60% with small increases in stock of 16%. This implies that more eggs per fish may have been produced at higher stock levels. If density-dependent fecundity held, the increment in fecundity would have been associated with a decline in stock. In general, Shelton and Armstrong (1983) have asserted that density dependent fecundity steepens the stock-recruitment growth curve.

Contradictions in research results about the stock-recruitment relationship have led to the belief that recruitment is determined by environmental factors. In addition to other factors, there are many errors in estimating the mortality of eggs and larvae (Cushing, 1988). However, there is no scientifically based reason to suggest that the density dependent function in stock-recruitment models does not occur (Cook and Armstrong, 1986).

The most important mechanism affecting growth, and consequently the fecundity and mortality of fish, is food availability to the larvae (Cushing, 1975). That includes also the availability of appropriate-sized planktonic organisms for larval consumption. Moreover, Garrod and Colebrook (1978) have attempted to establish a correlation between recruitment and environmental indices such as atmospheric pressure and climate. Although there are statistical deficiencies in this study, the conventional 0.05 level of probability that is used to judge significance may not be appropriate, bearing in mind the high variability of fish stocks (Garrod, 1977).

Other possible causes that affect fish growth can include the initial density of eggs (Daan, 1981), and diseases including viruses, parasites and stress. It is obvious to a fish biologist that the above factors are in many cases interrelated in the sense that one factor may be the result of another. In addition to the above, predation, and even cannibalism, (McCall, 1980; Santander *et al.*, 1983) can determine larvae and juvenile fish survival.

Overall, the nature of the stock-recruitment relationship is controversial throughout the scientific bibliography. This controversy sometimes arises from the fact that the existing stock-recruitment models take into consideration only a few of the many factors that affect growth, fecundity and mortality. Moreover, there may be factors applicable to single species models, which are extrapolated into the multiple species context without any pragmatic justification. In most stock-recruitment models the density dependent function is usually very simple. On the other hand, it is equally risky in terms of what happens in reality to use more detailed approaches to a function, where hope then takes the place of reality. Furthermore, in many cases the stock-recruitment curve is a hypothetical creation derived from a single fish stock statistical relationship, for which there is just enough evidence for a mathematical relationship to be formulated, and is then used to describe other fish stocks. Ricker (1973) and Hilborn (1985) discuss the stock recruitment relationship for stocks that consist of spatially separated substocks. As harvest rate increases, only these habitats that are more productive will maintain populations. In that case, a stock recruitment curve fixed in time is probably wrong. Instead, a bunch of different shape curves will arise depending on the state of exploitation.

From a fisheries manager's point of view, despite the above complications, stock-recruitment theory provides an insight into the most effective management approaches. In the case where stock is a good predictor of recruitment, the manager can manipulate stock to control recruitment to some desired end. The alternative and more realistic point of view is that there is not a straight forward density-dependent relationship between stock and recruitment. In that case, this knowledge indicates that effective management policies for improved exploitation patterns must incorporate uncertainty. Uncertainty about the fish resource may be reduced over time as a result of experience. This will be further discussed below.

Age-structured matrix models

Age-structured matrix models were developed from the Leslie matrix model (Leslie, 1948) and were widely used in ecology (Emlen, 1973) and demography (Keyfitz, 1968). Such models can be described by the general form

$$P_{k+1} = L \, P_k$$

where P_k is a vector of population size at age at time, k ie. $P_k = (P_k^1, P_k^2, ..., P_k^n)$ and the ith component, P_k^i, represents the number of individuals in the population of age i in life period k. The maximum age in the population is n years. The Leslie matrix, L, is an n x n matrix composed of age-specific fecundity f_i, and survival rates s_i, and has the form,

$$L = \begin{bmatrix} f_1 s_0 & f_2 s_0 & .... & f_n s_0 \\ s_1 & 0 & .... & 0 \\ 0 & s_2 & .... & 0 \\ .... & .... & & .... \\ 0 & 0 & .... & s_n \end{bmatrix}$$

Cohen and his colleagues (1983) applied the Leslie model to fisheries and they postulated that survival during the first year, s_0, is a random variable with specified mean and variance. Further, they computed analytical results for the average growth rate of fish population size as s_0 varied for the striped bass (*Morone saxatilis*). However, they considered that the coefficients f_i and s_i were independent of the population size. In this case, the Leslie model is linear and predicts exponential long-run growth or decline of the population. Generally, such approaches are of little pragmatic value for fisheries management because in reality, due to environmental and biological constraints, there is a combination of random effects and deterministic mechanisms that usually produce nonlinear models and determine the sustainable yields. Moreover, even in this simple density-independent model, Goodyear and Cristensen (1984) noted that the variability in s_0 obscured the parental stock effect.

On the other hand, density-dependent Leslie matrices, where fecundities f_i or survival s_i or both are assumed to be dependent on the population P_t, have been described by Getz (1984), Horwood and Shepherd (1981) and Reed (1983). However, such models, apart from their tendency to be complicated and their requirement of considerable mathematical sophistication, also exhibit several nonlinear multidimensional relationships that are seldom useful as an index of sensitivity of the fish population to noise, and the data requirements of these models seldom comply with available fishery data.

The Beverton-Holt Fisheries Model

In addition to the commonly used Schaefer population dynamic model (1957), another model with an application to fisheries management is the "dynamic pool" model of Beverton-Holt developed for the North Sea demersal fisheries by Beverton and Holt in 1957. The analysis of this model is facilitated by its attractive properties. Firstly, the parameters of the model have a clear biological interpretation and can be reasonably estimated from fishery data. Secondly, the model allows departure from and abandonment of the intriguing issue of the stock-recruitment relationship. Finally, the model is based on the effects of both fishing mortality and age of the first capture. Nevertheless, the model can be mainly seen only in a static yield-effort context. Beverton and Holt recognized this inadequacy of their techniques in determining the dynamic factors such as the optimal recovery of an overexploited stock. The dynamic analysis of this model is much more difficult than the dynamic analysis of the Schaefer model and the results that are obtained are many times incomplete.

The Beverton-Holt model presupposes that the fish population consists of a number of cohorts. Let $N_k(t)$ denote the number of fish belonging to the kth cohort alive at the time t, measured by calendar date. The time, k, corresponds to the time of recruitment of the kth cohort: that is the time that the cohort first becomes available to the fishing gear. Although it is unrealistic, recruitment is usually assumed to occur instantaneously at time k. Subsequently the time t-k, represents the age beyond recruitment and denotes the age of the cohort. The above can be expressed as,

$$N_k(t) = 0 \qquad \text{for } t < K$$
$$N_k(k) = R_k,$$

where R_k denotes the recruitment of the kth cohort.

The function $N_k(t)$ is assumed by the model to satisfy the differential equation,

$$\frac{N_k(t)}{dt} = - Z_{kt} N_k(t) \qquad (EQ.12)$$

where Z_{kt} defines the total mortality rate of the kth cohort and it is equal to $Z_{kt} = M + F_{kt}$. M represents the natural mortality rate and is assumed constant. F_{kt} represents the fishing mortality rate. It should be noted here that fishing mortality and natural mortality are independent of the stock size $N_k(t)$.

Integrating (eq. 12) we have,

$$N_k(t) = R_k \exp.\left[- \int_k^t (M + F_{kt}) \, dt \right] \qquad \text{for } t \geq k \qquad (EQ. 13)$$

By denoting w(a) as the average weight (kg) of a fish at age a, and using the most common parametric form in biology, the von Bertalanffy weight function we have,

$$w(a) = w_\infty \left(1 - be^{-ca} \right)^3 \qquad (EQ. 14)$$

where b and c are positive constants. This function is bounded, increasing and the weight approaches the asymptotic value w_∞.

The total biomass of the kth cohort at time t is given by,

$$B_k(t) = N_k(t)\, w(t-k) \quad \text{for any } t \geq k.$$

In the case where the fishing mortality rate, F_{kt}, is equal to zero, the "natural biomass", $B_{k0}(t)$ is equal to

$$B_{k0}(t) = R_k\, e^{-M(t-k)}\, w(t-k).$$

By differentiating it can be demonstrated that the natural biomass reaches a maximum at the age A (ie. T=k+A the critical age) which is determined by,

$$\frac{w'(A)}{w(A)} = M \qquad \text{(EQ. 15)}$$

Fisheries data on the values M and w_∞ and for the constants b and c (eq. 15) for many species of fish (Beverton, 1963; Beverton and Holt, 1959; Pauly and Murphy, 1982) have shown that the natural biomass has the shape as in figure 2.3. Figure 2.3 shows that $B_{k0}(t)$ first increases, then reaches a maximum at a point in time called the critical age and finally declines exponentially approaching zero at a maximum age as $t \to \infty$. Pauly and Murphy (1982) postulate that the natural mortality rate, M, for each species appears to be roughly specific to the family to which the fish belongs. For example, low natural mortality rates for flatfish of the family, Pleuronectidae, and high rates for the mackerel, Scombidae, imply that the average age of fish in an unfished population of the former might be about ten years and for the latter less than a year. However, a theoretically unrealistic conclusion deriving from eq. 15 is that the maximum possible yield from a given cohort could be achieved if the entire cohort were harvested at age T, by applying an infinite instant fishing mortality at that time.

In practice, in a fishery fishing effort is exerted at a finite value, so that a constant mortality coefficient, F, is applied to a given cohort. Following the Beverton-Holt model, suppose that a minimum mesh size, μ, is applied; theoretically this is a knife-edge selective measure in the sense that all fish of age $t-k \geq a_\mu$ that encounter the nets are captured and all smaller fish escape. Then, the total yield taken from the cohort can be calculated as a function of F and m by using the equation of the total biomass.

$$Y(F, a_\mu) = \int_{a_\mu + k}^{\infty} F\, N_k(t)\, w(t-k)\, dt$$

63

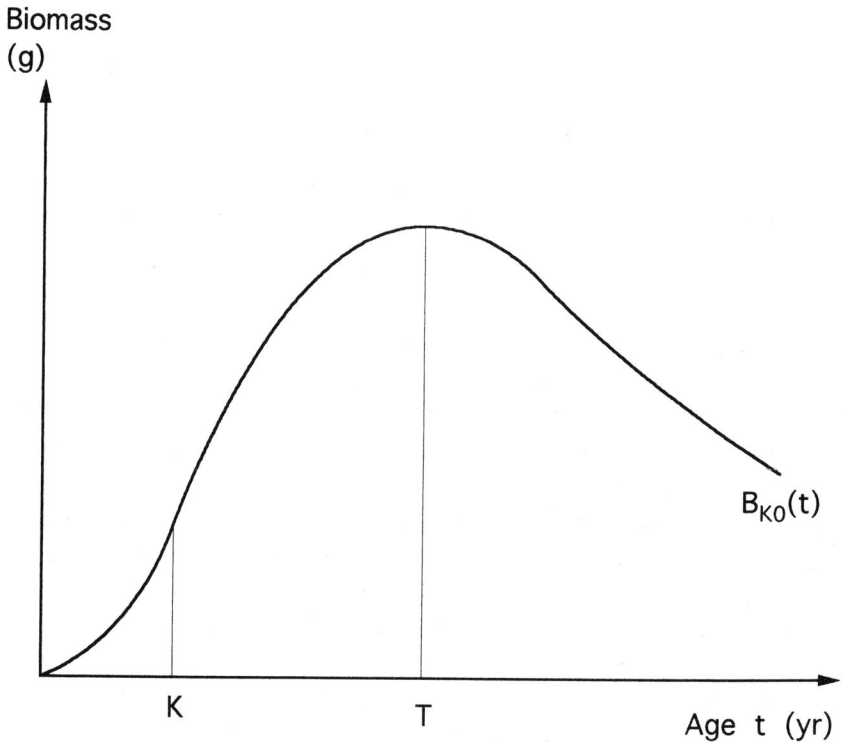

Figure 2.3   The Beverton-Holt fisheries model.
The natural biomass curve

Then, using eq. 13 the above equation becomes

$$Y(F,a_\mu) = F \, R_k \, e^{Fa\mu} \int_{a_\mu}^{\infty} e^{-Zu} \, w(u) \, du$$

if $w(u)$ is given explicitly by the von Bertalanffy weight function, then $Y(F,a_\mu)$ can also be calculated explicitly.

For any fixed level of F there exists some mesh size, μ, that results in the maximum possible sustained yield. The loci of all these points of the individual yield curves describe the maximum yield curve, called the eumetric yield curve (figure 2.4) and is defined by

$$Y_{eum.}(F) = \max_{a_\mu} Y(F,a_\mu)$$

Beverton and Holt (1957) show for North Sea plaice, that as the mesh size increases (for example in figure 2.4, $Y\mu3 > Y\mu2 > Y\mu1$) the corresponding peak in yield, that implies some fishing mortality rate F, ceases to exist. In other words, as the age of first capture rises the yield curve loses its peak and increases over the entire range of fishing mortality. Therefore, the yield curves suggest that concepts such as biological overfishing and maximum sustainable yield are inappropriate with regard to this model.

Clark and Kirkwood (1979) suggest a bioeconomic model for the management of the Australian prawn fishery located in the Gulf of Carpentaria based on the Beverton-Holt model. They consider "optimal" the eumetric yield curve $Y_{eum.}(F)$, since it represents the greatest sustainable yield that can be obtained at any given level of effort. Nevertheless, in a dynamic optimization problem with the application of discounting, the eumetric yield curve is of little significance. Therefore, the standard fishery production function, $Y=qEX$ in a static equilibrium analysis was used instead. Their results can be summarised in figure 2.5.

We can see from the figure 2.5, where average cost of fishing is assumed constant, that optimal fishing occurs at E^* where the net fishery revenue is maximized. The open-access equilibrium occurs at E_{max}. This simple bioeconomic analysis seems not to contribute to fisheries anything more than the Schaefer model does in a similar bioeconomic analysis. Indeed, it seems that the only distinction between the two models is the introduction of a mesh-size parameter in the Beverton-Holt model. Although the biological characteristics of certain fisheries such as the prawn fishery suggest the

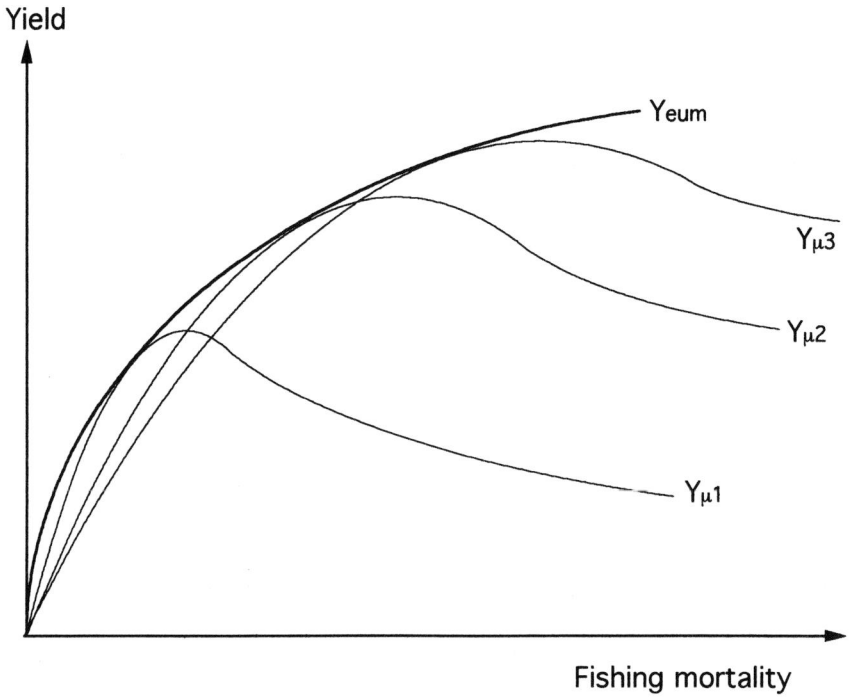

Figure 2.4   Eumetric yield curve as an envelope
of individual yield curves

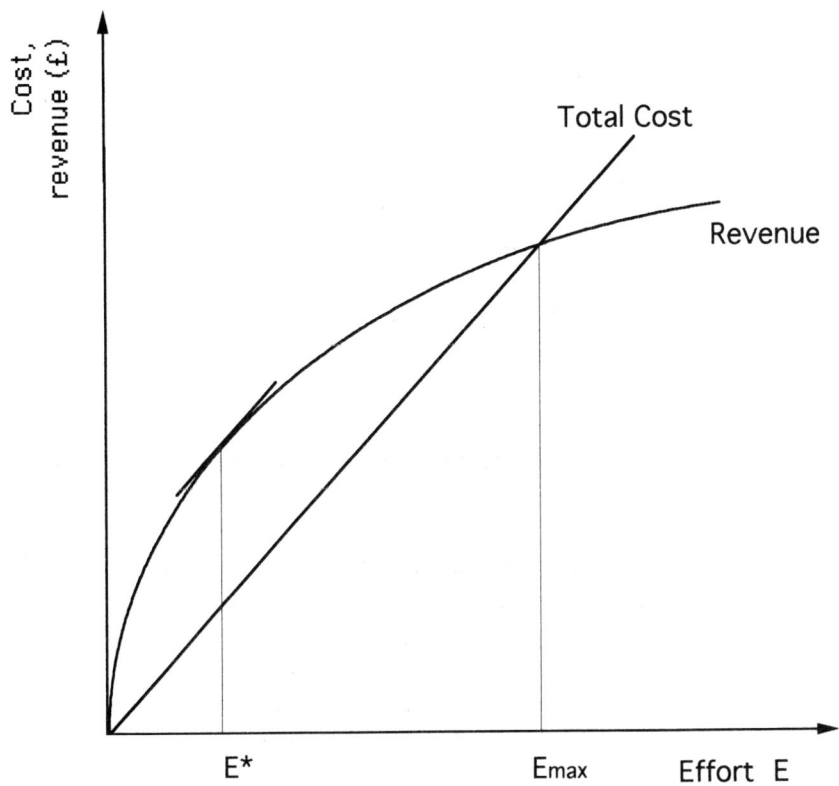

Figure 2.5   Static equilibrium analysis for
the Beverton-Holt model

inappropriateness of the Schaefer model, in a bioeconomic analysis it seems broadly justified.

The dynamic optimization problem for the Beverton-Holt model is studied by Clark and his colleagues (1973) using the data of Beverton and Holt (1957) for North Sea plaice. They found that discounting can have a strong effect on optimal harvest policy in terms of age of first capture and total fishing mortality. High discount rates induce heavier exploitation at earlier stages of the cohort's life. For example a 20% discount rate reduces the optimal age of first capture from 13.5 years to 7 years. In addition the optimal biomass level is not on the eumetric yield curve, which is considered a suboptimal level but on a lower "discounted" level. This can be reached more quickly but results in a reduced yield. Although Beverton and Holt recognized the need for a transitional phase in the recovery of an overexploited fishery, they could not determine an optimal recovery policy. In general, if a stock-recruitment relationship is to be included in dynamic analysis, the use of the Beverton-Holt model leads to results which are almost incomprehensible.

The role of stock assessment: Finding the best data representation model

The traditional view of stock assessment is a collection of analyses aimed at estimating stock size and productivity from statistics gathered from the commercial fishing process. In the case of the Common Fisheries Policy (CFP), catch and effort statistics are recorded in detailed logbooks for certain species and gear used in order to permit spatial mapping of relative fish abundance in the area that is fished. Then the catch is usually sampled for other characteristics of the fish, such as size and age composition.

The important role of the stock assessment is to help to distinguish whether catch and effort statistics are likely to give a misleading picture of stock trends and hence whether some more systematic and maybe expensive sampling programmes may be worthwhile. Such sampling programmes as research surveys, tagging experiments and tracking of individual fish, are planned and carried out by member state marine laboratories. Under such a scheme, the role of stock assessment is to provide the best possible technical support for a choice of policy options where the final decision is usually taken around a pivot of political, social and economic aspects.

The Schaefer dynamic model and its relatives are the most commonly used. They have formed the major assessment tool for several agencies for tuna and tropical fisheries and are widely applied in many temperate fisheries. These agencies include the International Convention for the Conservation of Atlantic Tunas (ICCAT), the International American Tropical Tuna Commission (IATTC) and the International Commission for South East Atlantic Fisheries (ICSEAF). In spite of the wide use of Schaefer based dynamic models, biologists in many circumstances look down upon these models and prefer to use the age-structured tools of Virtual Population Analysis (VPA) and Catch-at-age analysis (see below). Having previously surveyed the general principles of how fish populations behave and the various lines of thinking and mathematical notations used to describe different behavioural patterns, it is evident that in practice no single ideal model that attempts to represent the dynamics of exploited populations can cope with the variety of biological and ecological life history patterns and also with data that range from very simple and aggregated to very complex and detailed. The fact that there are no fixed biological parameters in reality that can be used to predict aspects of population behaviour, reduces the status of a model or an analysis to determining which simplifying convenience is good enough, or which is the best among several imperfect choices, for making predictions, in the face of the limited historical information available.

Among biologists there is often the misleading intuitive notion that the more detail there is, the better will be the population modelling. They assume that taking account of more aspects of population structure in space and time will eventually lead to more accurate predictions of population changes. There are two simple arguments against this notion:

Firstly, more detailed analyses usually lead to less accurate predictions due to cumulative errors, when many inaccurate parameter estimates are used.

Secondly, the effort and financial resources devoted to gathering detailed information can be redirected to other management activities such as enforcement and surveillance of existing regulations and measures that, for example, are desperately needed in the application of the Common Fisheries Policy (CFP). Hilborn and Walters (1992) argue from their experience that although it is often assumed that the "best model" forecasts short-term catches or stock sizes more accurately, such forecasts are not needed. This is because there is often an in-season regulatory system that provides good stock assessments and regulatory techniques within each fishing season, without having to rely on accurate predictions. Although this is a generalized argument, it shows emphatically how chaotic is the notion of "best model" and its implications for fisheries management.

In a theoretical work, O'Neil (1973) considered the problem of optimal model size in terms of measurement error, systematic bias and resulting model inaccuracy. Later, Walters (1986) considered the same problem in terms of

69

parameter uncertainty and prediction error. Both researchers suggested that there is an optimal model size (Figure 2.6). The optimal model complexity rises from the fact that as more variables are added, the ability of the model to fit the data will naturally improve. Nevertheless, the uncertainty and measurement errors in these parameters will increase even faster.

As mentioned above, between the two alternative methods for assessing the dynamic response of fish populations to exploitation, biologists usually choose the age-structured model and leave the Schaefer dynamic model as a second choice when age-structured data are unavailable. They claim that the main reason for not using Schaefer based models is their frequent failure to make sense of data sets, their "unrealistic simplicity" for the estimation of optimum effort and their equilibrium-based analysis of MSYs. In comparison, Hilborn (1979) showed that the so-called inefficiency of these models for the above reasons is not because of an inherent illogicality but to a poor quality of data on fishing effort and stock abundance. In addition, the same data failures plague the age-structured analysis. Consequently on this account Schaefer-based models surpass the complex age-structured models as they are straightforward and simple.

Recently, Ludwig and Walters (1985, 1989) tested the performance of age-structured models and dynamic models by means of simulations with artificially generated data series. They concluded that simpler models are likely to outperform the biologically more complex and "realistic" models because of the frequent difficulties associated with parameter estimations and the little informative variation in harvest patterns over time. Moreover, they state that they do not expect that additional more detailed information could compensate for the additional complexity that is required by age-structured models. Although such models seem more realistic and biologically correct, their requirements for complete uniform data can only be met in exceptional cases. Usually the information extracted from the available data ends up with parameters of "realistic" models, which are poorly estimated. The high cost of obtaining information about fish stocks condemns management strategies always to be developed in a twilight zone of inadequate information. As a consequence, under such conditions, fish stocks could be better managed by fitting the available data to the simpler Schaefer type dynamic models.

An extreme example is that in tropical fisheries, where an age-structured analysis is often impractical due to the difficulty of ageing the fish caught. In the case where the catch consists of many species, catch data are often impossible to collect for each species and management regulations are unrealistic if applied to each species. In these circumstances, it is more appropriate to apply the Schaefer model rather than treat each single species

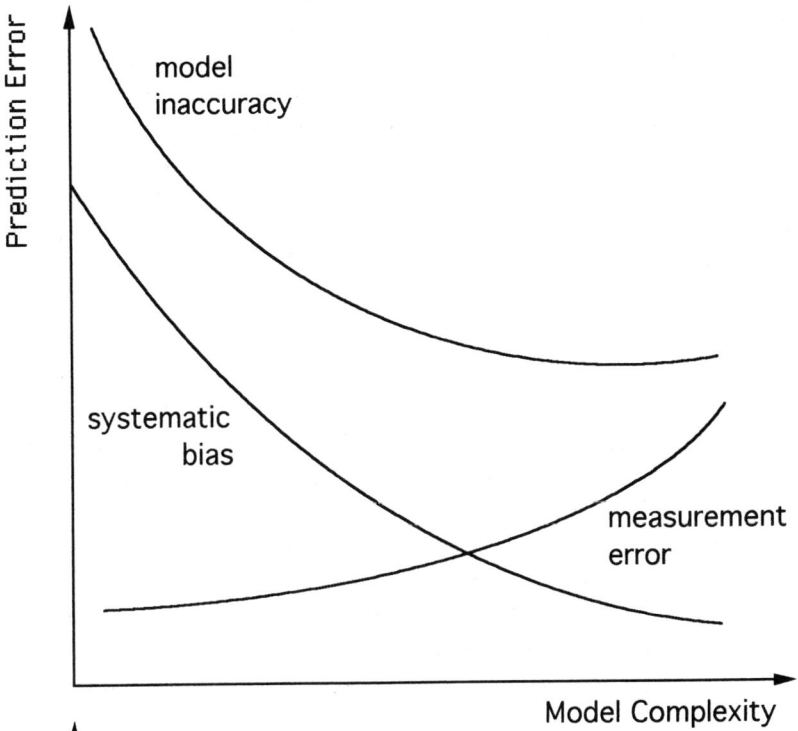

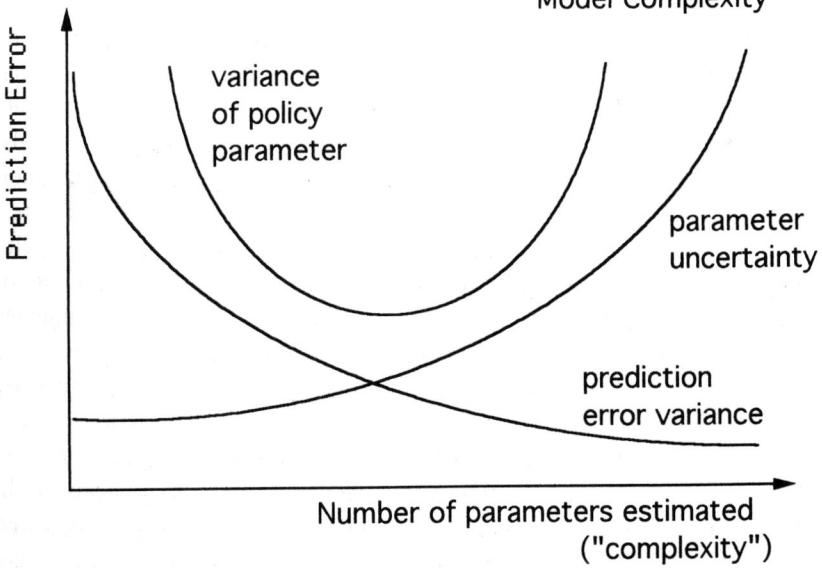

Figure 2.6 Optimal model size determinants

source: O'Neil (1973), Walters (1986)

separately in an age-structured analysis. With reference to a single species fishery, an example will be given for the halibut (*Hippoglussus stenolepis*), fished on the west coasts of Canada and the United States. The fishery is regulated by the International Pacific Halibut Commission (IPHC) since 1923. Halibut is fished using long line fishing gear and the data available to IPHC are catch, catch-at-age and fishing effort. The stock assessment performed by IPHC includes both types of analyses (Quinn *et al.*, 1985): firstly, there are age-structured analysis using the computer program CAGEAN and a similar analysis taking into consideration the movements of fish in the regulatory areas of IPHC and secondly, there is a biomass dynamic analysis, based on catch per unit effort (CPUE) data. The first analysis produce very good indices of what happened in the past but poor estimates of the current stock status. The second analysis based on CPUE data is used to determine the terminal fishing mortalities for the age-structured analysis. The failure of the management system is mainly due to incidental fishing mortality problems in the 1960s and 1970s that were not detected and incorporated into the stock assessments. As a consequence, the stock was inexplicably decreasing and restrictions in commercial harvest did not slow the decline as had been expected by fisheries managers and predicted by both analyses. In such cases, where the answers of both analyses came out the same and neither of them predicted the failure of the management system, the need for more detailed analysis can be strongly questioned.

To conclude, it could be suggested that when a fisheries policy decision-making authority decides on an array of applicable management factors and techniques by examining the socioeconomic aspects of the fishing industry, stock assessment scientists can make predictions or recommendations about the effectiveness of the management techniques on biological and ecological grounds.

The legitimacy of this task greatly depends on the relaxation of the usual demand of the policy-making authority that asks to be presented with definite predictions and answers from scientists and managers. It is easy to hire an "expert" to recommend what is wanted to be heard. Additionally, scientists have to make intelligent comparisons of alternative models. Such comparisons can be made by using the available fisheries data in several models and running simulations to decide which alternatives are most likely to give the most accurate results. Scientists have to explore deliberately and objectively a range of alternative models to make sure that the variations in mathematical notations of different models do not fool and obscure the presentation of the results. For example, when data show clear evidence of recruitment overfishing and the recruitment rate is an increasing function of the spawning stock, it does not really make any difference which equation is used, although many different equations and analyses describe the relationship between recruitment and spawning stock size. The policy prediction that the spawning stock should be allowed to increase is the same for all models.

When there is a difference in result among different models the scientists should try to understand the disagreements and analyze the management implications which arise from this difference rather than try to decide which model is absolutely correct. Fournier and Warburton (1989) showed that a composite estimation scheme using both age-structured and Schaefer type dynamic models outperforms either estimation model used by itself.

The role of Virtual Population Analysis (VPA) in producing results for fisheries decision making, a critique

The standard procedure in the North Atlantic at present is to use stock assessments that rely on catch-at-age data. These procedures are divided into two classes of models: i) the commonly called virtual population analysis (VPA) or cohort analysis and ii) the statistical catch-at-age methods. Both methods employ the historical record of catches and their age composition in order to reconstruct the initial size and subsequent history of each cohort that has already passed through the fishery. In that way, they form the basis for an age-structured catch forecast by employing a range of fishing mortalities of interest and finally they make propositions for the setting of TACs. A review of these methods is provided by Megrey (1989). The main difference between the two methods is that VPA uses recursive algorithms to derive its results while the statistical catch-at-age methods rely mainly on formal statistical models.

In general, VPA is preferred among the some twenty working groups of the Council for the Exploration of the Seas (ICES), which usually use different variations of VPA dealing with individual stocks or groups of species, and which meet regularly with Total Allowable Catch (TAC) calculations on the top of their agenda. The areas that ICES covers is shown in figure 2.7. Statistical catch-at-age methods are in general favoured among the North American scientists of the North Atlantic Fisheries Organization (NAFO), which covers the northwest Atlantic off the coasts of the USA, Canada and Greenland and which came into force in 1979. The European Community is a full member of the NAFO organization. Among other International Organizations, NAFO and, mainly, ICES provide the scientific input to the Commission's Directorate General (DG XIV) that along with the advisory Scientific and Technical

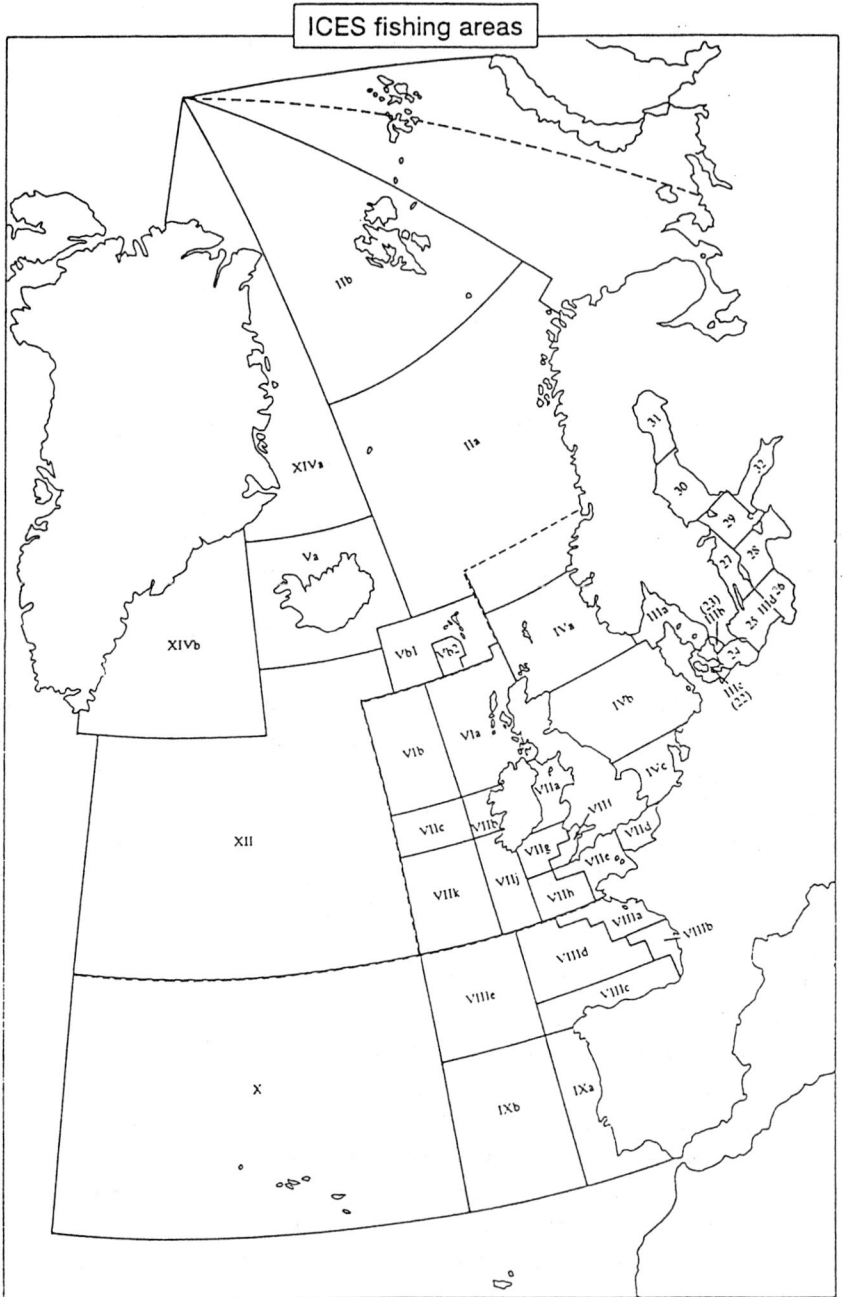

ICES fishing areas

Figure 2.7

74

Committees for Fisheries (STCF) analyses the information and puts forward TACs to the Council for approval.

In 1988 ICES held a workshop (Anon., 1988) in which the above methods were compared using data from simulation models. The results were not at all definitive. There were indications that the recent developments of methods called "tuning" the VPA (Anon., 1986a; Anon., 1987) work quite well and several of them consistently outperformed many of the statistical catch-at-age methods. As the objectives of this book are far from a detailed description of VPA and tuning techniques, it is briefly noted that tuning the VPA, using commmercial or research vessel survey CPUE data, is based on the relationship $F=qE$, where F is the instantaneous fishing mortality, q is the catchability coefficient and E is the fishing effort. Simply, it is the iterative procedure when the qs from complete cohorts are used to estimate Fs for the most recent year and to estimate qs for the incomplete cohorts. A variety of methods for tuning VPA are discussed by Laurec and Shepherd (1983) and Pope and Shepherd (1985). The main conclusion of the above workshop was that none of the methods concerned is clearly superior and more testing is needed in order to understand the merits of each of them.

Although VPA analysis is widely used, it is barely criticized. Hilborn and Walters (1992) consider that this is mainly due to the lack of independent estimates of stock sizes that can be compareed with estimates obtained by VPA. In an example given by them VPA consistently overestimates fish biomass. Another important reason in my view is that it is in generally hard to criticize an analysis which has many different variations and requires multiparameter procedures and techniques that are used by different ICES scientific groups. Moreover, VPA leads to results which depend on the idiosyncracy and experience of the scientific personnel that use it. However, in what follows, an attempt will be made to assess the major assumptions of VPA and their implications for fisheries management.

Terminal instantaneous Fishing mortality (F) and catchability (q) assumptions

One of the main problems of VPA is the assumption that there are no fish alive at a certain age, when commercial catch data are used. This is unavoidably determined by the selectivity of the fishing gear. It is often assumed that there are no fish older than the oldest fish ever to be caught. This is not true as any fishing gear is size selective. For example, hook and line gear is selective with respect to mouth size relative to hook size. The selectivity of trawling is related to mesh size, fish movement and gear avoidance. Furthermore, as fish grow they usually change habitats in order to avoid predators and to meet certain feeding and reproductive requirements. It is essential that additional surveys and tagging programmes, as well as detailed log books from commercial fishing, permit precise spatial location of catches. Stock distribution surveys should be conducted in order to find out a more representative age composition of a fish population. This problem of the terminal instantaneous fishing mortality (F) assumptions presents VPA with the difficulty that F estimations can only be made for cohorts that have passed completely through the fishery and not for cohorts that are still in the fishery and whose abundance is of great interest. The way that VPA deals with the problem is to make an independent estimate of the fishing mortality rate for cohorts that are currently being fished and to use this to estimate the cohort sizes relying on the equation,

$$N_t = \frac{C_t \ (F_t \ + \ M_t)}{F_t \left[1 \ - \ e^{-\left(F_t + M_t\right)} \right]}$$

where N_t is the cohort size at time t, C_t is the catch at time t and M_t is the natural mortality of the cohort. Usually in VPA F is estimated from effort data while assuming that q is known, by using the equation F=qE and relying on a constant q over time. This approach to the problem follows a procedure that was described earlier as the tuning of VPA.

However, catchability rarely remains a constant when CPUE data are used. This is due to possible changes in vessel characteristics (i.e. engine capacity, fishing gears), seasonality of fishery and differences in fishing areas. Although methods of analysis have been constructed, which add a trend to catchability they are rather dangerous to apply (Anon., 1986a). Various researchers, (Anon., 1987; Shepherd, 1988), stress that changes in catchability (q) are much less severe when research surveys of CPUE data are used. When shifts of q from aggregated commercial CPUE data occur, it is better to treat data from different resources and fleets separately. Saville and Bailey (1980) and Sinclair and his colleagues (1985) assert that the estimated biomass of Atlantic herring in the Northwest Atlantic using VPA during 1978-1980 is seriously overestimated. This problem usually arises when catchability, as estimated from VPA, has increased while the stock is declining but F has remained constantly high in the most recent years. In such cases, as in clupeoid fisheries (Ulltang, 1976) and in general in schooling fish stocks, catches tend to remain high although the stock is declining. Another example, where VPA failed for these reasons, is that of the Peruvian anchovy described by Csirke (1988). To overcome such problems with their major impact on the fishing industry, terminal F assumptions must be made in such a way that they reflect the anticipated changes in catchabilities. The need for an effective answer to this problem is usually avoided by imputing large cumulative fishing mortality to VPA as a standard procedure. Although this method has been considered beneficial in reducing errors created by arbitrary natural mortality assumptions (Sims, 1984), it is also known that they increased trends in F bias recruitment estimations obtained from VPA (Hilden, 1988).

Natural mortality (M) assumptions

Natural mortality is usually assumed known when it is imputed in VPA. However, it can be reliably estimated by the analysis of catch-at-age and effort data that allows for the correlation of total mortality with total effort (Shepherd, 1988). The intercept of such a relationship provides an estimate of natural mortality. The whole procedure is rarely applied, probably because it requires an extensive data set of high quality, substantial variation of effort in the data set and, in addition, tagging experiments. There are also methods of estimating M that are based on statistical models described by Pope and Stokes (1987). Overall, usually all methods that estimate M report large confidence intervals (Vetter, 1988).

The objective of this critique is not to review these methods but rather to elaborate on the implications of using in practice an assumed and often incorrect value of M in VPA. It should, however, be recognized here that since one often looks for trends rather than absolute numbers in population sizes, such biases for M may not be serious (Anon., 1986a; 1987). It is generally admitted that in cases of high rates of F, an incorrect M will have almost no impact on the estimations of stock abundance and recruitment. Nevertheless, recently LaPointe and his colleagues (1989) have shown that a period of high F, combined with overestimated M, produces deceptive VPA decreasing time trends in the estimated stock abundance and overestimations of the percent decline in biomass. Analogously, an underestimation of M leads to spurious increasing time trends. In general, even if the actual trend of F is known, an incorrect M value imputed in VPA produces deceptive changes that are severe for recruitment but less acute for total stock biomass. Similar results have been found by Hilden (1988) and by Sims, (1984).

Usually, although M is taken as constant in VPA this is hardly the case in reality. There are a lot of reasons why M might change over time, for example, changes in food availability, predators,and environmental factors. There has been an effort to include such effects in a variation of VPA called multispecies VPA (MSVPA) (Daan, 1987 and Pope and Knight, 1982). Indeed, in MSVPA the natural mortality rate for each species and each age depends upon the abundance of the species which preys upon it. Unfortunately, Daan (1987) who reviewed the North Sea MSVPA project concluded that the data requirements were unrealistic on this occasion.

Errors in age and size

As the whole foundation of VPA relies on catch-at-age time series, and indeed age composition data are treated by the analysis as though exact, the precision of such data is crucial. Although ageing techniques such as counting rings on otoliths, scales or other hard parts of fish are superior to age determination by length and weight they introduce new biases to estimation techniques such as VPA, which uses age data. What appears to be an annual increment (e.g. an otolith ring) does not necessarily mean that it is a truly annual increment and it could be attributed to spatial environmental perturbations or food availability (Beamish and McFarlane, 1983). Moreover, such ageing techniques are usually expensive, time consuming and sometimes impossible, especially for multispecies catches and tropical fisheries (see also previous section). Therefore, the questions to be asked are how precise must these data be and is there a threshold in error estimation, beyond which VPA is quite unreliable.

Let us assume that there is a 10% chance of misclassifying fish from a certain cohort as one year older than actual and a 10% chance of misclassifying it as one year younger. This misclassification produces a result of 200 fish wrongly assumed to belong to other age-classes out of 1000 evenly distributed at age individuals. This unrealistic example shows the magnitude of such errors, even when a 10% error in ageing is generally considered to be very good classification in practice (Hilborn and Walters 1992; Shepherd, 1988). Such misclassifications lead to unrealistic estimations of stock and recruitment sizes and are amplified when a strong cohort is followed by a weak cohort, coupled with a large M. In this case the age class one year behind the weak cohort would be estimated as much larger and with a lower mortality as well.

Similar biases would also affect the stock-recruitment relationship, where it may appear that a stock is much more productive at small stock sizes than it really is. These hypothetical examples show emphatically the detrimental effects of ageing errors on VPA. Although there are methods such as catch-at-age analysis with auxiliary information (see also below) they deal with such type of errors (Collie and Sissenwine, 1983; Deriso *et al.*, 1985), that are not widely used in practice.

Jones (1984) introduced another variation of VPA called length-based VPA that estimates recruitment and abundance-at-age by size distributions, which are easier to obtain. In a recent evaluation of this method Lai and Gallucci (1988) conclude that the sensitivity of this method to errors in the growth parameter estimations is considerable and it is critically dependent on having length frequency distributions. These distributions are only useful when they come from a population that is at equilibrium or when recruitment and exploitation rates have remained stable over time. Therefore, the results that length-based VPA produce are only valid for long term management advice, where equilibrium conditions can be achieved.

There is also a great difficulty in applying VPA to short lived species as for example in the case of the Peruvian anchovy (Csirke, 1988). The problem arises for such highly variable fish stocks bcause of the time-lag that is necessary for the estimation of the abundance-at-age. In that case monthly rather than annual data collection is needed for VPA. However, data gathering, such as catch and effort data, is usually much more time consuming than for a lifespan of a single cohort.

Stock identity assumptions

Another one of the main assumptions of VPA analysis is that there is no immigration or emigration from the stock. The reason for this is probably that in a short term prediction, VPA is able to assume that such mixing or separation of stocks is negligible. However, this short term prediction is based on past time series data where this assumption was also maintained throughout the analysis. These assumptions have major impacts on natural and fishing mortalities. If immigration into the stock occurs, then the number of fish will increase which will cause the cohort sizes to be overestimated. When immigration of fish of older ages takes place this will cause overestimation of the fish stock since VPA equations are solved backwards in time.

The impact of emigration is less severe when VPA is applied. This is because the effects of emigration can be incorporated into VPA as a form of natural mortality, especially when emigration is perfectly proportional to density (as M is). However, this incorporation depends greatly on how immediately VPA adapts to deal with changes in M, which, as has been shown above, is very poor.

Conclusions

In attempting to review the most striking deficiencies of VPA it can be concluded that the role of VPA as a forecasting procedure and advisory tool for fisheries management is rather ambiguous. Indeed, most of the scientific works referring to VPA usually pinpoint different unrealistic assumptions and deficiencies for VPA in analysing available data. Therefore, the multiparametric requirements of VPA, coupled with considerable uncertainty and the realistic complexity of the fisheries leads to the conclusions that i) the time trends of estimated abundance produced by VPA have an important role only in analysing and evaluating past management regulations based on the estimated response of the fish populations and ii) that it is to be hoped that in the future there will be a serious attempt to estimate the extent of these problems and correct VPA accordingly.

VPA, as it is applied nowadays, faces great difficulties with inaccurate biological parameters, and it is almost impossible for such a multiparametric recruitment analysis to deal effectively with other inaccuracies that exist mainly because of the current application of the CFP to the current confusing reality of the fishing industry. Most of the data that ICES working groups use in VPA rely on catch and effort from commercial fishing activities. Consequently, discarding, unreported catch, selling fish at sea or in different ports every time, incomplete logbooks and fishing induced mortality are essential problems that bring into question the reliability of VPA. Recent analytical and numerical analyses and techniques have shown that under commonly found circumstances such as the above, VPA outputs are quite sensitive to errors in inputs and assumptions (Rivard and Foy, 1987; Sampson, 1988).

Throughout the above discussion on the assumptions made by VPA it is evident that a variety of numerical and other sources of information (ie. different types of research surveys) to attain more accurate parameter estimation have been proposed. In addition, other numerical methods that deal with error measurements have been developed as varieties within the main body of VPA. However, they are not widely used in practice, mainly because they have expensive data gathering requirements and their results are still questionable and subject to further research. Therefore, in order to obtain more reliable estimations on stock assessment, more expenditure on research is necessary.

The problems with VPA such as those mentioned above have stimulated the development of nonsequential statistical fitting methods of estimating abundance (Fournier and Archibald, 1982; Gundmundsson, 1986) which now are gaining wider acceptance as plausible alternatives to VPA. The major contribution of these alternatives is the more formal way that they use data, for example in estimates of recruitment or fishing mortality and effort from sources other than catch in catch-at-age analysis. These methods are commonly called statistical catch-at-age analysis with auxiliary information and may form a powerful tool for future stock assessment analyses. However, most of the statistical catch-at-age analyses presently face other problems, for example terrible correlations between M and q due to a lack of systematic contrasts in effort levels over time (Hilborn and Walters, 1992).

In any case, for better decision-making it is very important for fisheries managers to be presented not only with the proposed future TACs estimated by VPA but also with the results of a sensitivity analysis as parameters are usually uncertain. Therefore, not only is the development of a more reliable stock assessment analytical tool essential but also much effort must be made by stock assessment scientists to present the non-technical fisheries decision-making authority with more readable results that show also the sensitivity and the uncertainty of key parameters used in stock assessment. This task is nowadays performed solely by biologists who usually rely on their idiosyncracies and their experience since there is not a formal scientific method to follow. Thus, scientific advice about fish populations is far from a consolidated scientific result and this should be stressed when such advice is used for decision-making in fisheries management. It is important that biologists acknowledge the limitations of their methods which should be clearly presented.

It is also essential for fisheries managers and for the decision-making authority to have a knowledge of the biological and ecological aspects of fish populations. Thus, they will be able to evaluate coherently scientific results and balance correctly the scientific information against the socioeconomic aspects of the fishing industry. Above all, a coherent understanding of the biological parameters that are imputed in the VPA may lead to better fisheries management decisions, as all these parameters also have great socioeconomic impacts for the fishing industry. For example, optimal catch interpreted as optimal F will have to vary according to the strength of the cohorts. In addition, uncertainty on future yield will probably lead to risk-averse decisions for building up a good market and receiving a better price for the fish.

3 Institutions of the European Community

The institution that is called the European Communities has its origin in the "European Coal and Steel Community" (ECSC) established in April 1951 by the Paris Treaty, the "European Atomic Energy Community" (Euratom) and the "European Economic Community", both set up in March 1957 by the Rome Treaty. These Communities had separate executive Commissions until 1965 when the institutions of the three separate E.C.s were merged and the Commission took over the powers and the responsibilities of all three Communities.

The basic institutional framework and co-operation procedure of the decision-making process is given briefly below. This framework takes into consideration the institutional reforms introduced by the "Single European Act", agreed in principle at the Luxemburg Summit (December 1985) and effective from July 1987. It is not the main scope of this work to refer in detail to these reforms; further information can be found in (Anon., 1986b; European, 1989; Noel, 1989).

The European Commission

The "College of Commissioners" comprise 17 Commissioners, two each from the UK, France, Italy, Spain and Germany and one each from the other states (i.e. Netherlands, Belgium, Luxemburg, Greece, Ireland, Denmark and Portugal). They are appointed by the state's government for a renewable period of four years. According to the oath taken on appointment they are expected to work as a single body pursuing the "Community interest". Each Commissioner has a personal staff (of about 15 people), called a "cabinet" which works directly with him and which helps him prepare his contribution to the work of the College and act as his contacts with the services of other institutions, with the Member States and with the general public. The Commission's President is elected by the College for a four year (renewable) period and enjoys high political status within and beyond the Community, ranking with Prime Ministers and Presidents of the Member States and represents the Community in the international arena. At present there are also six Vice Presidents.

The Commissioners are assisted by a total of 16300 people, who are the permanent staff of the Commission or temporary agents. 10000 of them are concerned with policy and executive tasks, about 3200 with scientific research, 2700 with language work, while 400 work in the Publications Office. The total cost of personnel and administration in the Commission budget represents some 3% of the total public spending of the Community, that is about 41 billion ECU and represents about 1% of the Community GDP. About 11500 people are grouped into 23 Directorates General (commonly known as D.G.s) each of which is responsible for a specific policy area (see table 3.1) for a list of DGs. The duties of the Commission are:

1   To act as a guardian and as an executive arm of the Community Treaties and legislation. If a breach of the Community Treaties, regulations or directives is made by any individual, organization or government, the Commission requires a reply within two months. If the Commission remains unconvinced or there is a failure to reply, then a "reasoned opinion" is issued and the body that contravenes must conform by a set date and/or pay a fine. If not, the Commission takes the subject to the European Court of Justice.

2   To be the initiator of Community policy and defender of the "Community interest". All Community proposals and legislation must originate in the Commission. A typical decision-making flow is as follows. Each Commissioner is assigned a portfolio to deal with a particular area of policy, for example the CFP. The D.G. responsible for this policy (for example DG XIV) is composed of junior officials and a chain of superiors, who prepare a draft proposal that is submitted to the "cabinet" of the responsible Commissioner and is to be approved by all the other DGs and other specialised services that might have an interest in the affair. Then the cabinet presents the draft proposal to the responsible Commissioner who asks the secretariat General of the Commission to submit it to the "college". The draft proposal must be presented in all the nine working languages of the Community. If the draft proposal is uncontroversial at the "college", it is adopted by the "written procedure" (normally in a week) or the "accelerated written procedure". If it is

Table 3.1 The European Commission's 23 Directorates-General and their specialised services

DG I	External Relations.
DG II	Economic and Financial affairs.
DG III	Internal Market and Industrial Affairs.
DG IV	Competition
DG V	Employment, Industrial Relationships and Social Affairs.
DG VI	Agriculture.
DG VII	Transport.
DG VIII	Development.
DG X	Information, Communication and Culture.
DG XI	Environment, Nuclear Safety and Civil Protection.
DG XII	Science, Research and Development; also the Joint Research Centre.
DG XIII	Telecommunications, Information Industries and Innovation.
DG XIV	Fisheries.
DG XV	Financial Institutions and Company Law.
DG XVI	Regional Policies.
DG XVII	Energy.
DG XVIII	Credit and Investments.
DG XIX	Budgets.
DG XX	Financial Control.
DG XXI	Customs Union and Indirect Taxation.
DG XXII	Co-ordination of Structural Policies.
DG XXIII	Enterprises' Policy, Distributive Trades, Tourism and Social Economy.

controversial, then it is rejected, modified, deferred or sent back to the DG. The "draft" proposal, when it is approved, becomes a "full" proposal and it is given a COMDOC number (Commission Document) and published in the Official Journal "C" series. This proposal is then referred to the Council Of Ministers which in turn submits it to the European Parliament and the Economic and Social Committee or to the various Community's committees and working groups of interest. All these bodies give their opinions and may suggest amendments.

3   To manage the structural funds i.e. the European Agricultural Guidance and Guarantee Fund (EAGGF), the European Regional Development Fund (ERDF), the European Development Fund (EDF) and the European Social Fund (ESF).

The Council of Ministers

This is the forum in which ministers from the Member States take the decisions. For example, Agricultural ministers sit in the Agriculture Council, Employment ministers in the Social Affairs Council, etc.; this type of Council is called the "technical" Council of ministers.

The Council's work is organised by the Committee of the Permanent Representatives (COREPER) established in 1958. The details of any proposed legislation are worked out by a number of small working groups made up of COREPER staff (in over 100 subcommittees and working groups). COREPER is divided into two main parts. The first part consists of the deputy permanent representatives of the member states, national civil servants or a mixture of them and usually deals with more technical and less theoretical aspects of the proposed legislation. The second part, consists of the permanent representatives of ambassadorial rank and deal with more political subjects. Usually, the "full" proposal is modified by the Council working groups in line with national interests and preoccupations. Then the two parts of COREPER try to reconcile the various interests seeking a Community minded solution and put a version before the Council of Ministers. The discussion in the "technical" Council takes a wider political character than in COREPER. Ministers frequently establish "links" with other issues, modify national positions, lift reservations or cajole colleagues. For matters of wide political importance or if there is a failure to reach agreement and the proposal is urgent, then the General Affairs Council is summoned. This consists of foreign ministers or the home secretaries of Member States.

The new co-operation procedure

One of the most important institutional provisions introduced by the Single European Act is the new co-operation procedure which substantially increases the legislative powers of the European Parliament. It introduces the cooperation between the three main Community institutions as follows. The Council must first adopt a "common position" by a qualified majority of votes. In the qualified majority vote, each Member State has a number of votes in proportion to its size, the UK, France, Germany and Italy have ten each, Spain has eight, Belgium, Greece, Portugal and Netherlands have five each, Denmark and Ireland three each and Luxemburg has two. 54 votes out of a total of 76 constitute a qualified majority. Although the majority voting principle was envisaged by the Rome Treaty and was to be applied from 1965, a severe crisis, resolved in 1966 by the so-called "Luxemburg Compromise", has meant that it has never come completely into effect. It was only used before the Single European Act, for technical and administrative decisions and for decisions relating to the budget and the Agricultural Management Committees, which had a specified time limit.

Then the Council sends the measure back to the Parliament for a second reading. The Parliament must within three months approve, amend or reject the Council's "common position". This can only be by the absolute majority of the European members of Parliament i.e. 260 votes, half the number of seats (518) plus one. Then it goes back to the Commission, which has one month to revise the proposal, and finally back to the Council. The Council has three months to do one of the following

1 adopt, by qualified majority the revised Commission proposal;
2 adopt, by unanimity, Parliament's amendments not approved by the Commission;
3 amend, and then adopt, by unanimity the Commission's proposals;
4 ignore the revised proposal and adopt by unanimity the original proposal'
5 fail to act, in which case the whole procedure starts from the begining.

In addition, the Commission may at any time before the Council has adopted the proposal, amend or withdraw it.

The European Parliament

The Parliament consists of 518 MEP (Members of the European Parliament) chosen through direct elections in the 12 Member States for a period of five years. The four most populous Member States i.e. France, the UK, Italy and Germany each have 81 MEP, Spain has 60, the Netherlands 25, Belgium, Greece and Portugal 24, Denmark 16, Ireland 15 and Luxemburg 6. Since German unification 18 observers from the eastern Länder have been taking part in Parliament's proceedings. The MEP divide their time between their Brussels offices and their constituencies, political group meetings and plenary sessions in Strasbourg of one week per month, where often intensive lobbying of MEP by interest groups takes place. The role of Parliament is mainly consultative and advisory rather than legislative, but is gradually acquiring more power in the decision-making process. The power of the European Parliament has been enhanced to a considerable degree by the Single European Act. The second reading resolution (mentioned above), that is required by the Single European Act, must be approved by the majority of the full membership of the Parliament and not just by the majority of those present at a debate. Although a final decision still rests with the Council and the Commission, which has the power to amend or withdraw its proposal, the Parliament now has the ability to send a powerful message to the Commission and the Council. In the case where the Commission accepts the Parliament's amendments, then together they can exert considerable pressure on the Council, since it can only change the revised proposal by unanimity.

The European Court of Justice

The Court sits in Luxemburg and comprises 13 judges assisted by six advocates-general that are appointed for a six-year term by common agreement between the Member States. It has two main functions:

1   It inspects the laws enacted by the Council and ensures that they are compatible with the Treaties. Such cases can be brought before the Court by E.C. institutions, by a Member State or by members of the public.
2   It pronounces on the correct interpretation of or the validity of Community provisions when requested by a national court.

It has played a decisive role in the shaping of the Community law and its rulings take precedence over those of national courts in this area.

The Court of Auditors

This is the fifth Community institution according to the status given by the Maastricht Treaty. It was set up on 22 July 1975 by a treaty that came into force on 1 June 1977. It has 9 members, increased to 12, under the Maastricht Treaty. Its members are appointed by the Council for a six-year term. Its role is to check "in a lawful and regular manner" that revenue is received, that expenditure is incurred and that the Community's financial affairs are properly managed. An example of the role of the Court of Auditors in the fisheries sector is that in its Third Annual Report it pointed out that much of the aid under the Regulation 17/64 went to the construction and modernisation of vessels for herring fishing. Nevertheless, at the same time the Community was imposing significant restrictions on herring fishing for conservation reasons. As a result, many of the vessels receiving Community aid had very limited fishing opportunities [22]. Such findings are set out in annual reports, at the end of each financial year.

A general overview of the decision-making process and of the E.C. institutions is given in figure 3.1. In general, as happens in cases relevant to fisheries, there is not a single rigid process by which legislation is adopted. The Community's legislative instruments are usually "regulations" "decisions" and "directives". The regulations and decisions apply directly to all Member States and the directives require national implementing legislation. The adopted regulation is published in the Official Journal "L" series. In addition to the above institutions there is a plethora (about 700) of advisory committees and pressure groups organised under the E.C. umbrella.

The most important advisory body, composed of 189 representatives, is the Economic and Social Committee (ECOSOC). Its members fall into three main groups, union officials, employers' representatives and people concerned with more general public interests, for example professional bodies, consumers, environmentalists etc. ECOSOC is consulted on a large range of policy matters by the Commission and also publishes its own detailed reports and recommendations on its own initiative. During 1958-61 the Council was obliged to consult the ECOSOC on legislation relevant to the CFP and on the common organisation of the market in fisheries products based on article 43 of the Rome Treaty.

Among the various groups and committees there are a number which have been specifically set up to advise the Commission in the field of fisheries:

1   The Advisory Committee on Fisheries established by the Commission in 1971 [4]. It consists of 45 members from different sectors of the fishing

88

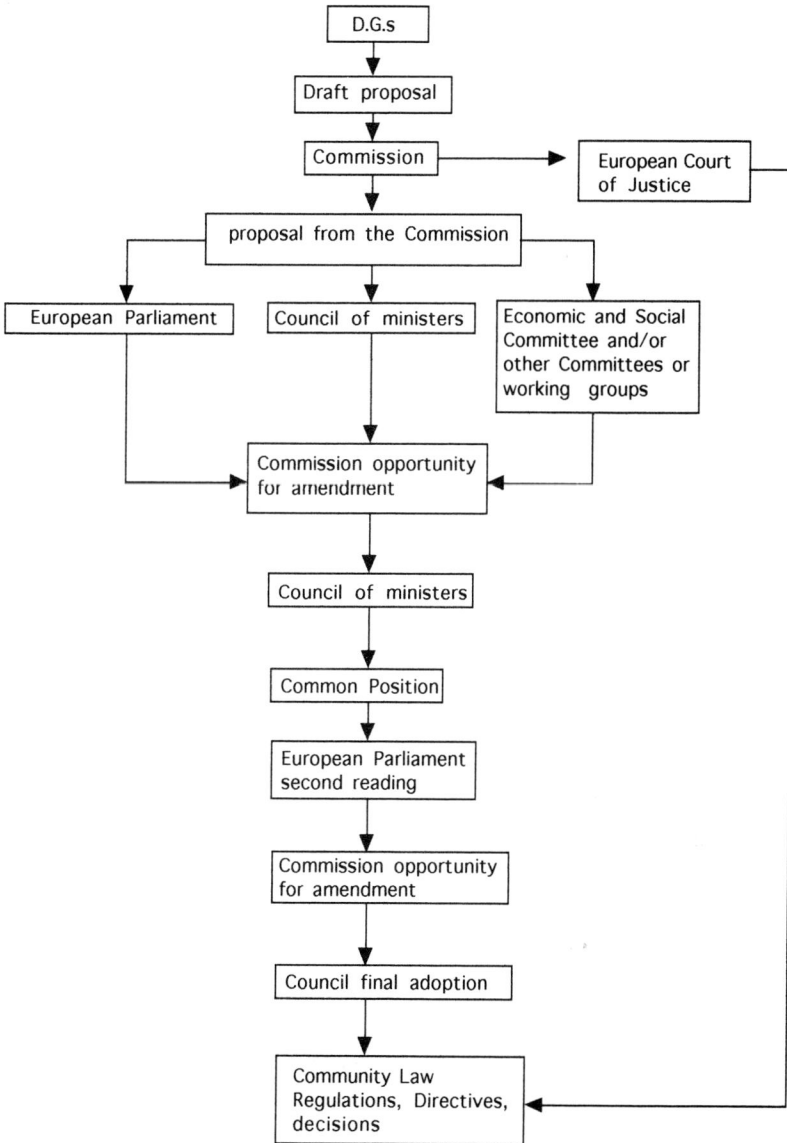

```
            ┌─────────────┐
            │    D.G.s    │
            └─────────────┘
                   │
                   ▼
          ┌───────────────┐
          │ Draft proposal│
          └───────────────┘
                   │
                   ▼
          ┌───────────────┐          ┌─────────────────┐
          │  Commission   │─────────▶│ European Court  │
          └───────────────┘          │   of Justice    │
                   │                  └─────────────────┘
                   ▼
      ┌───────────────────────────┐
      │ proposal from the Commission│
      └───────────────────────────┘
```

Figure 3.1   E.C. Institutions and co-operation procedure

89

industry, for example, traders, representatives of the fish processing industry, credit institutions, producers and workers. Its role [5] is to give opinions on "any problem concerning the operation of the Regulations and measures relating to the establishment of a common structural policy in fisheries or to the common organisation of the market in fishery products ... and also on all social problems arising in the industry, with the exception of industrial disputes"

2   The Joint Committee on Social Problems in Sea Fishing established by the Commission in 1974 [6] to replace the Joint Advisory Committee on Social Questions arising in the Sea Fishing Industry set up in 1968 [7]. Its 44 members are representatives of employers' and employees' associations. The Committee gives its opinions to the Commission on "the formulation and implementation of the Community social policy aimed at improving and harmonising the living and working conditions in sea fishing" on its own initiative or at the request of the Commission. Furthermore, this Committee was set up to "promote dialogue and conciliation and facilitate negotiations between certain listed employers and employees' associations" [8].

3   The Scientific and Technical Committee for fisheries (STCF) plays a prominent role on fisheries matters by drawing up annual reports with regard to fishery resources, conserving fishing grounds and stocks (art. 12 (1)). It was first set up by the Commission in 1979 [2] and then amended by the Third Act of Accession in order to enlarge the membership of the Committee [3]. It consists of 28 fishery scientists. This has been replaced by the Scientific, Technical and Economic Committee for Fisheries (STECF), established at the end of 1992 under the new fisheries regulation [170]. The Committee is to be consulted on the biological, technical and economic factors concerning the fishery resource. It also has to produce an annual report on the situation of the resource with reference to the above general factors.

4   The Management Committee for Fishery Products, originally set up in 1970 [18], subsequently amended in 1981 [19] and recently in 1992 [72], covers aspects of marketing and trade in the fisheries sector. It consists of Member States' representatives with a representative of the Commission acting as chairman. The Committee gives its opinion on the Commission's draft proposals by a qualified majority of votes weighed proportionally to Member States.

5   The Management Committee for Fisheries and Aquaculture, established by the new fisheries regulation in the end of 1992 [170], has replaced the Management Committee for Fishery Resources, established in 1983 [20]

to give its opinion on measures proposed by the Commission to conserve stocks. The Committee's functions and procedure follow the usual pattern above.

6   The Standing Committee for the Fishing Industry set up in 1970 [21] "to promote the co-ordination of structural policies for the fishing industry and to ensure close and constant co-operation between Member States and the Commission". The Committee's functions and procedures follow the usual pattern described above.

Conclusions

In this chapter, an overview of the European Community institutions and their functions has been offered. Although the organisation appears to be bureaucratic and has been criticized for being that, it offers a comprehensive framework which could enable effective policy formation. It is apparent that the formation of policies that would cater for the often conflicting interests of several different countries and groups represent a giant task. The development of the European Community has often been affected by nationalism, economic recession and enlargement of the Community. Until the Single European Act unblocked the Community's decision-making process, the Community has been preoccupied with the subject of institutional reforms and its implications for European integration.

In recent years, the significant changes which have occurred have served to consolidate and strengthen the legal basis, the institutional framework and the policy-competence of the European Community. However, the continuing importance of external pressures or national political and economic interests, which continue to determine the limits of European integration, should be stressed.

4 The legislative framework of the CFP

The Treaty of Rome (25 March 1957)

The treaty of Rome demanded and established in member states the adoption of a common policy in the sphere of agriculture (article. 3(d)). Later, in art. 38 agricultural products are defined as "the products of the soil, of stock farming and of fisheries and products of first stage processing directly related to these products". Within the framework of the overall Common Agricultural Policy (CAP), article 39(1) sets out the objectives that by definition (article. 38 above) could be shared by the Common Fisheries Policy (CFP). These are:

i To increase productivity by promoting technical progress and to ensure rational development and the optimum utilization of factors of production.
ii To ensure a fair standard of living for the agricultural community.
iii To stabilize markets
iv To ensure the availability of supplies for consumers at reasonable prices.
   In order to attain these objectives the Member States are obliged by art. 5 to "take all the appropriate measures" while ensuring "no discrimination on grounds of nationality" (art.7).

First legislative acts of the Council

The Commission constructed a package of proposals in 1966 [9] in response to conflicts in fish trade raised by differences in individual duties, the Community's Common Customs Tariff and the progressive liberalisation within the framework of the OECD (Organization for Economic Co-operation and Development) and GATT agreements. These proposals were made mainly

92

for the benefit of French interests to avoid clashes over trade in fish. They took a final form in the first Council regulations on an integrated CFP [10,11] applied from 1 February 1971. A considerable effort was made by the Council to guarantee the objectives of the Rome Treaty when formulating these regulations. These regulations established firstly a common organisation of the market in fishery products that supported prices and protected and conformed to the Community market (see also below). Secondly, a Community Structural Policy was set up that aimed to co-ordinate the structural policies of Member States (see also below). Thirdly, the basic principles of the Rome Treaty were underlined once more, for example, the principle of freedom of access to community fishing grounds and the complete ban on national discrimination between community citizens. Finally, a concern about conservation and overfishing in the fishing zones of Member States was expressed in a call for the more "rational" exploitation of resources without defining exactly what "rational" meant.

The 1972 Act of Accession

The final agreement between the original six Member States and the three new Member States (UK, Denmark and Ireland) was incorporated into the Treaty or Acts of Accession. The Acts reflected the highly developed fishery interests of the new Member States (whose aggregate catch was twice that of the six Member States) with specific and important references to the fisheries sector.

Under articles 98 and 99, Guide Prices covered certain fishery products and the common organisation of the market was reinforced. Under articles 100 to 103 the fishing rights of Member States were determined. Specifically a temporary derogation from the principle of freedom of access to waters of other Member States was established by defining certain areas, where Member States could extend the six nautical miles of their coastal fishing areas to 12 nautical miles. This derogation was in force until 31 December 1982 when the Commission had to present to the Council a report, on which the latter would examine what provisions might supersede the provisions of the Act of Accession. These 12-mile fishing zones are shown in figure 4.1. In art. 102 the Council was directed from the beginning of 1978 to decide on measures to regulate fishing effort for protecting and conserving the fish resources.

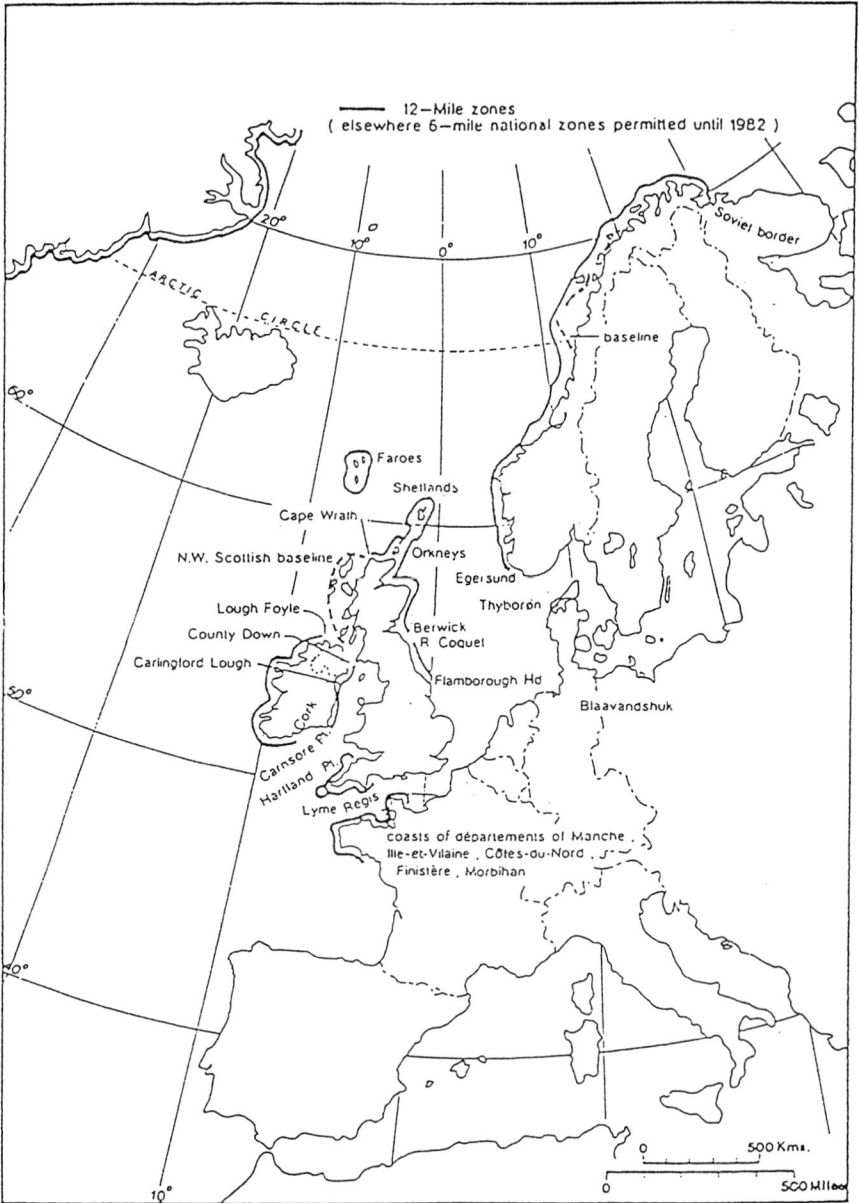

Figure 4.1 The 12-mile fishing zones under the E.C. Treaty of Accession

source: From Wise, M. (1984). *The CFP of the Europen Community* London, N.Y.: Methuen & Co Ltd.

The creation of 200-Miles EEZ, towards a comprehensive CFP

The most significant turning point in the worldwide development of fisheries, and one which also accelerated the need for a more comprehensive CFP, was the declaration of exclusive economic zones (EEZ) by a number of nations as well as by the E.C. on 3 November 1976 in The Hague. The claim to property rights over seas as in EEZs was a gradual phenomenon worldwide that reflected a change in thinking and in the interests of governments over the exploitation of natural resources, including fisheries.

In successive United Nations Conferences on the Law of the Sea (UNCLOS), beginning 1958 and specifically in UNCLOS III, which started in Caracas in 1974, a key issue was the extension of EEZ. At the end of the UNCLOS III session in May 1975 most participants accepted that coastal states should have jurisdiction over the exploitation of natural resources within EEZ. In Europe the gradual extension of the Icelandic exclusive fishing limit from 4 miles in 1952 to 200 miles in October 1975 won general recognition, despite UK opposition. Then Norway followed on 1 January 1977. In the rest of the world some countries, for example, Chile, Peru and Ecuador had extended their zone in 1952 long before the first UNCLOS. By 1974, 38 states had extended their jurisdiction beyond 12 miles; among them were India and many South American, African and South East Asian states. Moreover, Mexico in 1975, Australia, China and Canada in January 1977 and the USA in March 1977 extended their jurisdiction to 200-miles in order to exploit natural resources.

In response to this, the Member States agreed to extend their fishing limits to 200 miles (370 km) from their North Sea and Atlantic coasts in a resolution passed on 3 November 1976 in The Hague in the Foreign Affairs Council. This came into force on 1 January 1977 (not published in the Official Journal, however a summary was published in 1981) [15]. In this way a legal spatial framework finally developed with Community regulations to conserve stocks and allocate them among Member States and this framework was based to a certain degree on previous Commission proposals. These Commission proposals for a conservation policy [12] aimed at "an optimum yield from the resources". The proposed basic instrument for this objective was the fixing of an annual Total Allowable Catch (TAC) for each stock or group of stocks and the decision to "distribute the permissible catches fairly amongst the fishermen of Member States, using a system of quotas". Furthermore, the Foreign ministers agreed that certain regions were "particularly dependent" on fishing and special "account should be taken of their vital needs".

The same Hague Resolutions with regard to the CFP were further elaborated and developed by the declaration of 30 May 1980 by the Foreign Affairs

Council [16]. The Council agreed that a common overall fisheries policy in compliance with the Hague Resolutions should come into effect on 1 January 1981. This Council declaration contained guidelines for "the adoption of a structural measure which includes a financial contribution by the Community" and for the "establishment of a securely based fisheries relation with third countries". Furthermore, the principle of "relative stability" i.e. the maintenance of a fixed quota per stock for each Member State of the resources available to E.C. [17] was enunciated for the first time taking into account the following factors:

1   In an attempt to take account of the traditional fishing patterns of the Community fleet, the catches during the period 1973-1978 were used as a point of reference.

2   Certain preferences (later termed the "Hague preferences") were to be applied to fishermen in certain regions where there were (and still are) few job opportunities for alternative employment. The regions that enjoy "Hague preferences" were Ireland, Northern Ireland, the Isle of Man, Scotland, the North-East coast of England from Bridlington to Berwick and Greenland. Quotas for Ireland were set at twice as much as the 1975 catches. For the Northern regions of the UK clarified by the 1980 proposals [23], the Commission proposed to guarantee "local fishermen" a minimum allocation equal to their catches in 1975. The selection of this year was believed to have met the "essential needs" of such fishermen taken from the North Sea and the West of Scotland. The term "local fishermen" was defined to apply to those that use boats under 24m. in length and operate out of "local ports dependent mainly of fishing".

3   The losses suffered by Member States' fishing vessels in third countries after the introduction of 200-mile EEZ have also been considered.

In its first attempt to apply the May 1980 criteria the Commission [13] based its proposed quota allocations for the seven major species (ie. cod, haddock, saithe, whiting, plaice, redfish and mackerel) on cod equivalents. The calculation was achieved by giving the value of one to cod, haddock and plaice, 0.87 to redfish, 0.86 to whiting, 0.77 to saithe and later 0.3 to mackerel. Table 4.1 shows that the calculation of the quota allocation in tonnes of cod equivalent for the seven main species followed three steps. As a first step the average catch in 1973-78 in E.C. waters and the catch possibility for 1981 in non-E.C. waters was estimated for each Member State. Then, under the "Hague preferences", the catch in E.C waters of the first step was compensated

Table 4.1   Proposed quota allocation in tonnes of cod equivalent for the seven major species

	E.E.C.	W. Germany	France	Nether-lands	Belgium	U.K. North*	U.K. South	Denmark Mainland	Denmark Greenland*	Ireland*
Catch possibilities for 1981 in EC waters based on average catches during 1973-8, plus catch possibilities for 1981 in non-EC waters	1,120,643	156,767	159,636	99,827	26,525	204,879	139,569	268,650	43,151	21,639
	100%	13.99%	14.25%	8.91%	2.37%	18.28%	12.45%	23.97%	3.85%	1.93%
Catch possibilities for 1981 in EC waters where Member States contribute to "Hague Regions" under the "Hague preferences" plus catch possibilities in non-EC waters for 1981	1,120,643 (±44,224)	148,577 (-8,190)	151,589 (-8,047)	94,612 (-5,215)	25,139 (-1,386)	215,460 (+10,581)	132,277 (-7,292)	254,614 (-14,036) (+11,764)	54,915	43,460 (+21,821)
	100%	13.26%	13.53%	8.44%	2.24%	19.23%	11.8%	22.72%	4.9%	3.88%
Compensation transfers among Member States for losses of catch in third country waters based on 1973-76 average catches	-	+36,22	-25,749	-19,780	-5,049	-	+75,789	-61,430	-	-
Total catch possibilities	1,120,643	184,797	125,840	74,832	20,090	215,460	208,066	193,184	54,915	43,460
Total % of EC	100%	16.49%	11.23%	6.68%	1.79%	19.23%	18.57%	17.24%	4.9%	3.88%

* "Hague Regions", specially dependent upon fishing regions according to the "Hague Resolutions"

source: SEC(1981) 105 (12.6.1980) and O.J. (1982) C22/2 (29.1.1982) (numbers in brackets) by catch transfers so that Member States made a contribution for the specially dependent regions. As the third step the losses of catch in third country waters were estimated based on the 1973-76 average catches, and a compensation mechanism followed by transferring catches between Member States that aimed to equalise these losses after the introduction of 200-miles EEZ. Finally as a result of the above, the quota allocations for 1981 for each Member State were calculated [24, 25].

In 1982 the UNCLOS convention in Jamaica found 93 countries that had already established or were about to proclaim 200-mile zones. This represented the equivalent of 35% of the ocean area that contained 95% of all the marine living resources. The article 61-64 of this convention dealt with the rational use of fishery resources for conservation and management. Up to now over 120 states including the Community (in 1984) have signed these provisions.

Before the introduction of 200-mile EEZ the E.C. attempted to assess the impact of such a measure on Member States' fishing fleets. In a 1975 report [14] based on data for 1973 the Commission calculated that about 72% of the Community catch would fall within the 200-mile EEZ around the E.C.. The above figure however, is misleading if it is taken by itself, because Member States generally made catches of high value species in the waters of third countries, and these are usually worth much more than the catches in Community waters. For example from the same report [14] it can be estimated that, although losses of cod in tonnes are 41% of the total gross losses, in terms of money (EUA) the cod losses represent 59% but such estimations must be treated sceptically. Furthermore the whole 1975 report is somewhat unrealistic in that it is based on the extreme assumption that a total ban on Community fishermen would be applied by non-E.C. countries and that there would be a total exclusion from E.C. "pond" of non-E.C. vessels. The almost catastrophic and unrealistic conclusions that can be drawn from such reports usually give additional arguments to Euro-sceptics to use their favourite words "cumbersome and bureaucratic" when they refer to E.C. institutions.

Nevertheless, the Community's response to EEZ was immediate. Negotiations started with non-E.C. countries under four major types of agreements:

1   Reciprocal agreements with countries that have interlinked fishing interests (e.g.. Norway, Iceland, Sweden).

2   Agreements based on access to surplus resources (e.g. USA).

3   Agreements based, in principal, on mutual access to resource and markets for both parties (e.g. Canada).

4   Agreements based on the acquisition of fishing rights in third country waters with compensation that can be financial, trading concessions, the exchange of fishing rights, training, or a combination of some of these elements (e.g. Senegal, Guinea Bissau, Angola).

5   Agreements based on a combination of the previous two types (access to stocks/access to markets/financial compensation) (e.g. Morocco, Greenland)

The Council resolution towards a coherent single CFP package

The Council resolution of 25 January 1983 is comparable in many respects with the CFP resolution concluded in the Hague in 1976. However, the new policy of 1983 combines various new and existing measures in a single and coherent CFP package. In principle it has been applied until the 31 December 1992 when the Commission had to produce a report on the situation in the fisheries sector, on the economic and social trends in coastal regions and the state of resources. On the basis of this report it will be decided if any adjustment should be made according to the procedure laid down in the article 43 of the Rome Treaty before the policy is extended until the 31 December 2002 [26]. In fact, the new Council regulation of 1992 [170] made some generic adjustments to the 1983 resolution and establish a Community system for fisheries and aquaculture which came in power in the 1.1.1993. The 1983 resolutions remain the spine of the CFP for all the specific measures for fisheries which appear also in the new regulation. There are some new elements that are introduced which are presented in most generic form in the new regulation and may apply after 1994 until the 31 December 2002.

There are three regulations and one declaration that comprise the key aspects of the Council agreement of 25 January 1983:

1   The establishment of a Community system for the conservation and management of fishery resources [27].

2   The fixing of TACs for 1982 for certain fish stocks, their quota allocation between the Member States and the conditions under which TACs may be fished [28].

3   The establishment of certain technical measures for the conservation of fishery resources [29].

4   A declaration of intent in the form of a Council Resolution on the implementation of a structural policy in the fisheries sector (within the following six months) [30].

99

The new 1992 regulation [170] relies on the above provisions and declares the following new elements without any specific definition:

1   The establishment of a Community system of licences for Community's fishing vessels that is to be applied no later than 1995. Member States will operate this scheme as a national system of fishing licences. However, what kind of licences they are to be and what information they will contain that will be attached to a vessel is not specified.

2   The Council may establish Total Allowable Catches (TACs) on a multiannual (MATAC) and/or on multi-species (MSTAC) basis after careful consideration on a case-by-case examination. Detailed rules for the application of the Council's intention are to be expected sometime in the future. The main specific elements of the 1983 resolutions as amended by the new regulation of 1992 are as follows.

Restriction of access

The derogation of the freedom of access established in the 1972 act of Accession was maintained and also the former six-mile limit was extended to 12 miles around all Member States. However, historic rights within 6 to 12 miles must be respected. Preliminary negotiations determined the precise extent in the 6-12 mile zone of the traditional-historical rights of other Member States to be maintained over a ten year period with the possibility of "adjustments" after a ten year period according to articles 6 and 14(2) of the 1992 regulation [170] to ensure "relative stability".

The principle of "relative stability" continued to apply and was incorporated in the article 4(1) [27]. The double role of the "relative stability" principle and particularly the "Hague preferences" as a socioeconomic as well as a conservation measure to protect the biologically sensitive areas, justifies the establishment of the Shetland/Orkney "Box" round these islands. The Community also introduced a licensing scheme for certain fishing activities restricting the rights of large vessels (over 26m) to enter this "Box", on 29 July 1983 [31], and this is also adopted in the 1992 regulation [170]. Restrictions were placed on the catching of fish for human consumption and no fishing is allowed for industrial stocks. This licensing scheme is administered directly by the Commission.

The table below (table 4.2) shows the fishing activity in the Shetland Box. This table is constructed from data by the Scottish Office Department that record the presence of individual vessels in the Box at some stage during the year. Therefore, these figures may not represent any substantial degree of fishing effort, which may be considerably lower than the above statistics imply. For example, 1988 appears to be the year of the highest fishing activity with 3(UK, 40 French, 8 German and 2 Belgian vessels within the Box; however, some of these vessels may not have been involved in any fishing activity in the Box. The permitted number of licences are 62 for the UK, 62 for France, 12 for Germany and 2 for Belgium. It is obvious from this table that a considerable underutilization of the fishing opportunities in this area occurs and it can be asserted that the licence scheme is not an effective measure for restricting the fishing activity in this area. Hence, probably the most beneficial conservation effect of this measure is that vessels from Holland and Denmark, that usually are involved in industrial fishing, were not allocated licences to fish in the Box

Before continuing the description and analysis of the remaining legislative aspects of the CFP, one should mention the problem of the Euro-style in which the regulations are written and which does nothing to facilitate their study. A non exhaustible list may start with the cumbersome and unnecessary cross-referencing. One of the many examples which appears in important Community's regulations (such as in regulations 3760/92 establishing a Community system for fisheries and aquaculture and in 3759/92 for the Common Organization of the market) is that almost at the end of each article appears the statement "detailed rules for the application of this article shall be adopted in accordance with the procedure laid down in article" 18 for the first regulation and 32 for the second. Articles 18 or 32 are identical. A new cross-reference specifies that one should refer back to article 148(2) of the Rome Treaty, in order to elucidate the procedure, for example how representatives' votes shall be weighted. Furthermore, the innovative linguistic acrobatics evident in every regulation could be justified only in artistic writing. It is not the scope of this book to explore this subject. Maybe the imprecision and vagueness of specific measures and notions, which probably stem from difficult political compromises, are assorted by the language which treats such compromises in a broad sweeping way. However, such a judgement is difficult to justify and maybe the recent opening of the doors of the Council of Ministers to television (2.2.1993) will show it.

Table 4.2   Fishing Activity in the "Shetland Box", number of vessels by country during the years 1983-1990

	United Kingdom	France	Germany	Belgium
1983	-	51	1	0
1984	8	55	1	0
1985	8	54	0	0
1986	22	49	6	0
1987	36	37	7	1
1988	36	40	8	2
1989	32	41	12	1
1990	30	43	7	1
permitted number of vessels under the licensing scheme	62	52	12	2

Source: "Memorandum by the Fishermens' Association, the Orkney Fisheries Association, the Shetland Islands Council, and the Orkney Islands Council" in Review of the Common Fisheries Policy, House of Lords Select Commitee on the E.C. 23.6.92 and Reg. 3760/92 in O.J. (31.12.1992) L 389/1

The Community's practice of setting annual Total Allowable Catches (TACs), first proposed in 1976 [12], was legally established by article 3 of the E.C. regulation 170/83 [27]. The Scientific and Technical Committee for Fisheries, set up by article 12 of the above regulation, was to play an important role in the formulation of TACs based "in the light of the available scientific advice". In the process of the establishing of TACs for stocks shared between Member States and third countries and also in negotiations for access to stocks of non-E.C. countries, the E.C. would act as a single body. Then the TACs are to be distributed to Member States according to article 4 [27] taking into consideration the three key factors of the "relative stability" principle mentioned above. The distribution of 1982 quotas, according to the Commission's proposal in 1982 [32] was to be the "reference allocation" or the "status quo" for the future allocations that had to observe the principle of "relative stability". For quota allocations the calculations were made in terms of "cod equivalent" and the stocks were divided into four categories as follows:

i) The seven major species (cod, haddock, saithe, whiting, plaice, redfish and mackerel-later herring replaced redfish)

ii) Other species for human consumption (pollack, hake, sole, megrin, anglerfish, Norway lobster, penaeus shrimp).

iii) Herring, although the majority of herring fisheries were closed at the time.

iv) Industrial fish, i.e. Norway pout, sprat, horse mackerel, blue whiting.

Despite the balance achieved with the reference allocation problems have arisen in subsequent years which still have not found a definite solution. In theory, the initial "reference allocation" had to be made resilient and adaptable to changes in fish stock populations and to economic parameters of the fishing industry. However, since a political compromise was involved, the application of the principle of "relative stability" is far from being a precise, static, mathematical relationship, as it was meant to be. In addition, the allocations of 1983 concerned the ten Member States at that time. While German unification in 1990 did not affect the fishing opportunities of the other Member States, two new members (Spain and Portugal) have disputed certain aspects of the way the principle is applied and they have initiated proceedings before the European Court of Justice against the quota allocations among the Member States for 1990, 1991 and 1992 (cases 7,63,67.70,71,73,129/90-1; 84-86,99,151/91) in the waters of Greenland, the Faroe Isles, Norway and Sweden. Some of these cases are pending but the Court, in a judgement of 13 October 1992 for the cases 7,63,67,70 and 73/90, has totally rejected some of them. However, the Court recognised that the two Member States can apply to be included in the distribution of new fishing possibilities with non-E.C. countries and in any eventual review of the distribution system in accordance with article 4(2) of the E.C. regulation 170/83 [27].

Technical and conservation measures

The 1983 CFP aims, as the title of the Regulation 170/83 [27] states, to "establish a Community system for the conservation and management of fishery resources". In order to meet this objective the regulation establishes the Advisory Scientific and Technical Committee for Fisheries under the auspices of the Commission (article 12) for consultation at "regular intervals". Under the 1992 regulation [170] this Committee is replaced by the Scientific, Technical and Economic Committee for fisheries. The Committee has to draw up an annual report on the situation of fishery resources, with reference to the developments in fishing activity in relation to biological and technical factors. For the first time the Commission envisaged that the Committee has to take into consideration "the economic implications of the fishery resources situation". The meaning and the effect of this economic appraisal is to be seen in the future. In addition, a Management Committee for Fisheries and Aquaculture, consisting of representatives from the Member States under the chairmanship of the Commission, has been set up by article 17 of the 1992 regulation. The advice of the former Committee to the Commission is also based on scientific inputs from the reports of the Working Groups of ICES channelled through its Advisory Committee for Fisheries Management (ACFM). Moreover, other scientific input comes from the North Atlantic Fisheries Organisation (NAFO) established in 1979 and from the Northeast Atlantic Fisheries Convention (NEAFC), which came into force in March 1982 and of which the E.C. is a full member. Both the above international organisations replaced former organisations after the introduction of the 200-mile EEZ and they make recommendations and give advice on fish stocks in the Community waters as well as on fisheries under reciprocal agreements with third countries, for example the USA and Canada.

In addition to TACs, the 1983 CFP adopted a comprehensive set of technical conservation measures, under the Regulation 171/83 [29], of indefinite duration. This regulation adopted a batch of measures, whose technical complexity can be summarised as follows. It prescribed minimum mesh size, maximum permitted by-catch levels and minimum fish sizes to be landed. It also prohibited fishing for salmon beyond 12-mile limits in most areas and established close seasons and areas for redfish and herring to protect spawning grounds and nursery areas. Furthermore, it prohibited the use of certain types of gear and vessels in certain areas for certain species. Such technical measures owe their introduction to scientific advice. For example, minimum mesh sizes aim to prevent the catching of immature fish and allow stocks to grow. By-catch regulations seek to limit the amount of human consumption species taken in fishing for industrial species.

In addition to the above measures Member States in certain circumstances are allowed to take conservation measures by themselves. Under article 18 of the above 1983 regulation (exactly like its predecessor article 19 of the Regulation 2527/80 (33)), Member States may take appropriate measures when the conservation of certain fish stocks or fishing grounds is seriously threatened and where any delay would result in damage which could be difficult to repair. In such a case the Commission must be informed, then confirm, cancel or amend such a measure and the Council may then amend the Commission's decision. Under the 1983 regulation 171/83, article 20, which has been re-stated in article 10 of the 1992 regulation (170) (there is no equivalent article in the Regulation 2527/80) a Member State may adopt measures that involve strictly local stocks and are applicable only to its native fishermen. Such measures are intended firstly to ensure better management and better use of fish, crustaceans and molluscs in such a way as to take into account "the needs of both producers and consumers". Secondly, they aim to establish special measures, not provided for in this regulation for fisheries and areas concerned, to ensure that non-commercial activities do not jeopardize the conservation of the resources. The above measures have been upheld by the European Court of Justice in cases 804/79, 87/82 and 86-87/84, where it asserted that a Member State has not only the right to maintain certain technical measures but also has a duty, where it is necessary for action to be taken to meet an established conservation need. In the latter cases (87/82 and 86-87/84) the Court accepted that the Regulation 171/83 contained a number of lacunae and asserted that, until the Community takes the necessary action, it is in the Member State's interest to introduce legislation to allow fishing to proceed in an orderly fashion and to operate consistently with the aims of the Community legislation. Such national legislation might also be justified under the article 20 of the 1983 regulation 171/83. Indeed, after the adoption of this regulation there have been numerous national technical measures. From 1983-85 there were 224 new measures, and 512 pre-existing local measures were notified to the Commission as required by articles 19 and 20 (39).

Although the Regulation 171/83 was prepared and debated by the Commission over a long period of time after 1980, and subsequently although the European Court declared certain lacunae in this regulation, the Commission did not adopt the necessary amendments until the middle of 1984. These amendments included for example, how mesh sizes are to be measured, permitted exceptions and prohibitions on attaching obstructing devices to nets [34] and laying down a sampling procedure for measuring by-catches [35].

Moreover in addition to the inactivity of the Commission (the Commission also blamed the Council's procrastination over the adoption of increased minimum mesh sizes [36]) the competence of Regulation 171/83 was also limited to Community waters (including the waters of Portugal and Spain) but not to the Baltic and Mediterranean. In June 1986 a more comprehensive technical conservation regulation than that of 1983 was adopted for the Baltic [37]. This Baltic Regulation was more self-contained than that of 1983 without the lacunae described above. It was based on the Commission's previous experience and on the recommendations of the International Baltic Sea Fisheries Commission.

The Common Fisheries Policy and the Mediterranean

The Mediterranean fisheries account for about 15% of total fish production and 30% of total value in the Community. Although the stocks comprise valuable species, it is difficult to give a precise estimate of the stock situation, since the demersal stocks, which are the most intensively harvested, consist of various species with different life spans and complicated biological characteristics. Consequently, the situation of the stocks is difficult to model. Generally it has been estimated that demersal stocks are fully exploited, while stocks of small pelagic fish (sardine and anchovy) are exploited to a moderate extent. The situation becomes further complicated by the lack of reliable catch figures and the fact that the same stocks are exploited by different fishing fleets.

As far as the fishing industry is concerned, this is divided into two distinct categories: 1) The inshore fishing fleet which is characterized by a large number (some 47,000) of small-boats, under-powered and undecked of less than 9 metres involved in "family type" fishing with a high number of low-paid jobs. This is usually characterized by the old age of the craft and the equipment and the lack of appropriate safety conditions on board. 2) The high-sea fishing fleet which operates in areas outside territorial waters (6 or 12 miles offshore, 200 miles zones do not apply in the Mediterranean) using relatively modern vessels, which catch tuna and other large migratory species.

The Common Fisheries Policy, as defined in 1983, applies to the Mediterranean only in part: Only the rules on market organization and structures are fully applicable, whereas those on conservation and management of the resources are not. It was not until 1990 that the Commission issued a number of proposals for the improved management of Mediterranean fisheries [187]. This resulted in a Council regulation of 1991, which is still in existence [188]. The regulation establishes a Community framework for studies and pilot projects relating to the conservation and management of fishery resources in the Mediterranean. The main priority areas are the development of the structure of traditional fisheries and specialized fisheries, such as sponge, coral, sea-urchin and seaweed. Additionally, it introduced the monitoring of fishing activities via the development of a statistical network, the coordination of research and the use of scientific data. Recently, (1993), the Commission submitted a proposal which aims to co-ordinate the variety of technical measures in the Mediterranean [189]. These include the standardization of fishing gear, minimum mesh size, minimum allowable fish size and areas in which fishing is allowed only at a certain distance from the coast or at a particular depth.

The proposal of 1990 [187] suggests that the responsibilty for the implementation and administration of the policies concerned should be entrusted to bodies, which would enjoy powers delegated by the public authorities. These would be composed of the various trade and professional interests. It is assumed that this will result in collective responsibility, which will make the producers comply with the rules. The proposal suggests an active involvement of producers in the implementation of the policies. Experience elsewhere of decisions of fishermen's organizations with their emphasis on present revenues makes it likely that this policy will not conserve the resource. The procedure is also likely to leave itself open to abuse by politicians of a formal or informal nature.

It is apparent that fisheries management in the Mediterranean exists at an embryonic stage and it cannot be postulated what effect the regulations will have. It would be advisable, however, that the E.C. should drop uniformity in regulations, since the fisheries concerned are determined by various individual characteristics and are based to a great extent on traditional practices which will be difficult to modify. The species concerned have not been adequately studied for biologists to confirm particular trends, and this is one of the areas to which the Community has paid particular attention. Nevertheless, the funding provided has been widely criticized as inadequate [167].

5 The CFP instruments

The competence of the CFP and its instruments fall into three main categories:

i) resource management and conservation policy,
ii) common organisation of the market, marketing policy and
iii) structural policy, monitoring and surveillance.

The budgetary management for fisheries and its relative importance in the economy

The expenditure flow on the Community's fisheries and its breakdown among the particular policies of the CFP is shown in figure 5.1. In 1993, a total sum of 618.5 million ECUs was committed by the Community for fisheries, representing an increase of 25.5% compared with that of 1992. Most of the Community's expenditure on fisheries is spent on the structural policy. In 1993 this represents 63% of the total (figure 5.2) fisheries expenditure followed by the fisheries agreements (25%) and the Common organization of the market (5%). Overall, the total Community's commitment to fisheries represents less than 1% of the Community's total expenditure (68,000 million ECUs), which in turn comprises only 1.2% of the GDP of the twelve Member States in 1993 [169,171]. The Community's expenditure on fisheries is extremely small, when compared with that for Agriculture (35,000 million ECU) which represents 49% of the total Community expenditure. E.C. fisheries expenditure in 1993 is comparable with that for food, aid and support operations of the Community for Third World countries [169,171].

Nevertheless, the full amount of the commitment appropriations is not utilized by the Member States. For example, the Community estimates, according to the

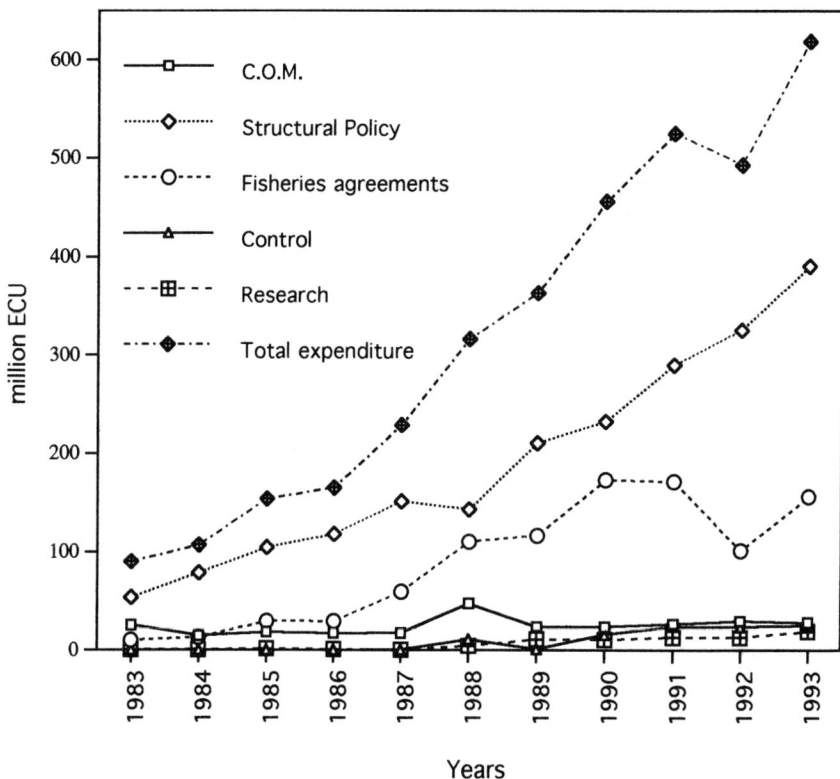

note: Data1983-1985 for 10 Member States, data 1986-1993 for 12
Member States. C.O.M.: Common Organization of the Market

Figure 5.1   Trend of Comminity's commitment
               appropriations for fisheries during
               1983 to 1993

source: SEC(91) 2288 (18.12.1991) and General Budget of EC in O.J.
(8.2.1993) L31

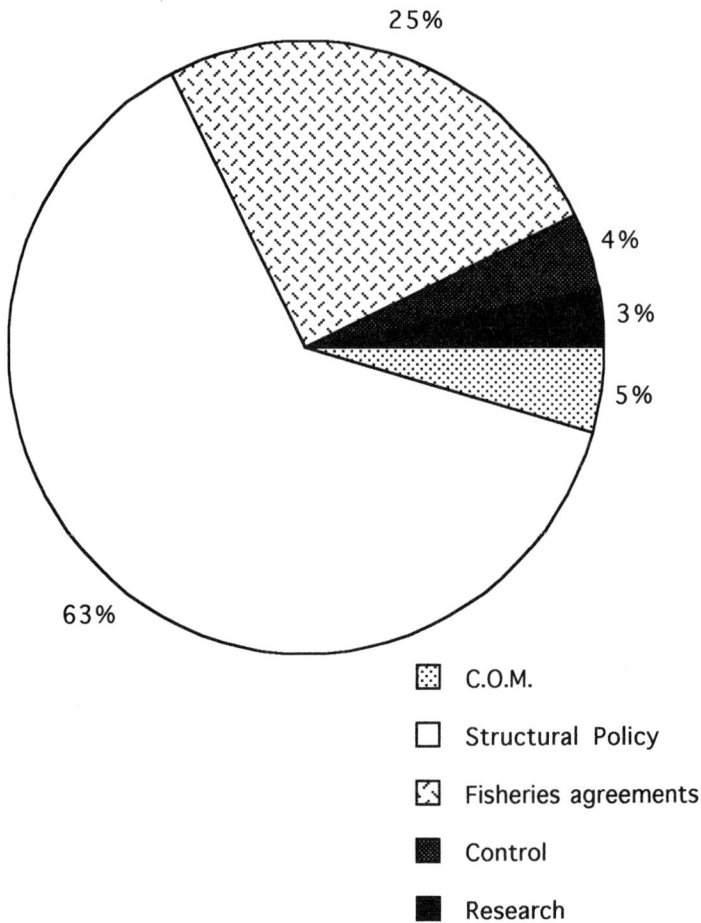

25%

4%

3%

5%

63%

▨ C.O.M.

☐ Structural Policy

◩ Fisheries agreements

◼ Control

◼ Research

Figure 5.2 Allocation of Community's
commitment appropriations for
fisheries in 1993

source: Final adoption of the general budget of the E.C. for the financial
year 1993 in O.J. (8.2.1993)

applications that have been received, that by the end of 1993 a sum of 415 million ECUs will have been utilized out of the total commitment appropriation for fisheries of 618.5 million ECUs, a total which represents 0.6% of the Community's total expenditure for this year. There is a substantial difference between the total amount of commitments and payments for fisheries for the years 1987 to 1993 (figure 5.3). This is attributed to the unused funds for measures under the structural policy of the CFP as well as for marine inspection and surveillance controls. However, under a Council Decision of 1988 [175] the unutilized amount in a given year can either be allocated to Member States or repossessed by the Community during any subsequent financial year depending on the date when the application is submitted.

In aggregate economic terms the relative importance of the fishing industry in the Community's economy seems also of relatively small importance. For example, the value of landings of the eleven Member States as a percentage of the E.C. G.D.P. remains about 0.15% [148] since the last Community enlargement in 1986. Greece and Portugal have the greatest proportion of more than 1% of their G.D.P. Moreover, the number of fishermen as a percentage of the national workforce in the Community was less than 0.4% in 1990 [54,148]. Again, Greece and Portugal have the greatest proportion of more than 1% of their national workforce. Nevertheless, such crude economic indicators may misleadingly underestimate the importance of the fishing industry. The socioeconomic importance of fishing in certain coastal regions, where fishing is a localized primary activity with little possibility for alternative employment, has been underlined in the so-called "Hague Preferences", amended by the 1986 structural regulation [146] to include coastal regions in South and North Spain and the whole of Portugal.

In addition to the regional importance of fishing activities for the local population there is a series of activities related to fisheries of importance to the economy and hidden under the above economic indicators. The main industries related to fisheries are aquaculture, the processing industry and auxiliary industries.

For example, in 1989 the Community's aquaculture produced a total of 825,234 tonnes of fish, molluscs and crustaceans with a turnover value of 1,400 million ECUs [167], which represents about 20% of the total landing value of fish caught at sea. The Community's expenditure on aquaculture projects during the years 1983-1991 was 225 million ECUs [54,171] i.e. about 16% of the Community's structural funds for the fishing industry over the same period. Unfortunately, there are no available data in the Community to show how many people are employed in this business. Although this book does not aim to analyse the merits or disadvantages of the aquaculture industry,

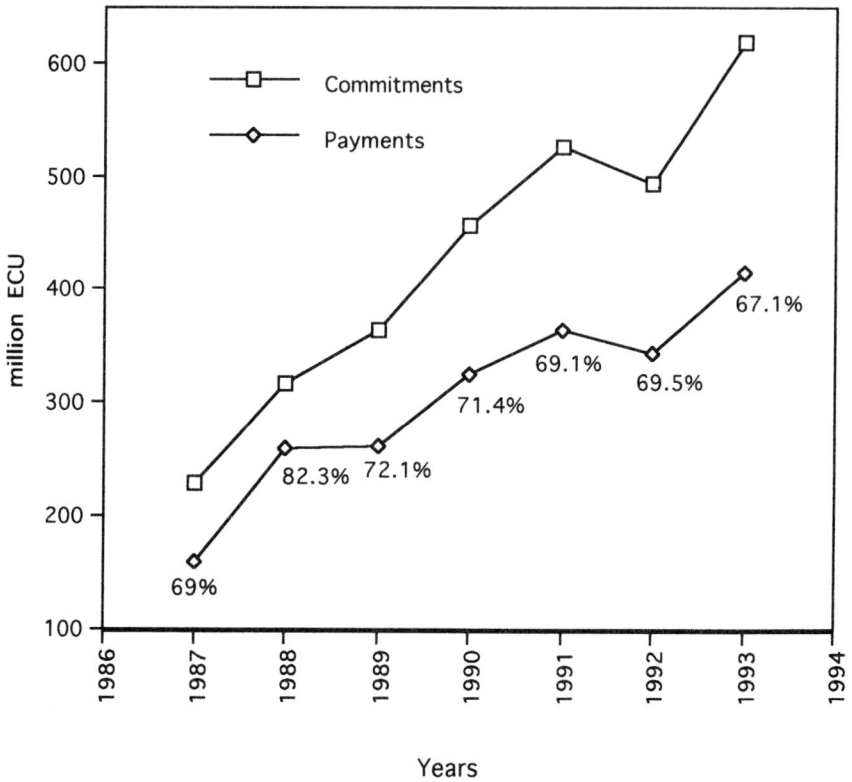

Figure 5.3 Trend of Community's total
commitments and payments for
fisheries and their rate of utilization
for each year during 1987 to 1993

source: Data from O.J. (15.2.1993) C330 and General Budget of
the E.C. for the financial year 1993 in O.J. (8.2.1993) L31

it is unquestionable that aquaculture has been a good source of alternative employment mainly in areas where fishing is the traditional source of employment. While the press often presents aquaculture as a business that has reached its peak in many places in the Community including the West of Scotland and Greece, the situation may be more promising if one considers that out of a total of 866 aquaculture projects financed by the Community during 1983 to 1989, the success rate was about 95% [167,172] and that the Community still continues to offer grants for aquaculture projects [172].

Regarding the data available for the processing industry for fisheries products in Member States, these draw the most varied and uncertain picture with reference to the relative importance of this industry. Under a 1989 E.C. regulation [173] for the improvement of the processing and marketing of fishery products the Member States had to draw up a sectoral plan for their countries and establish the situation of the processing industry. This was to act as a guideline for the "Community support framework 1991-1993" for the industry. Despite this request, which should have resulted in definitive statistics, only six Member States gave definite information about their processing industry. As far as the structure of the workforce is concerned, the ratio of employees in the processing industry varies considerably among the Member States. For example, in 1989, whereas in Denmark they represent 58% of the total working force in the fishing industry in Greece they comprise only 6% [174,125].

The picture becomes more heterogeneous where the Community canning industry for fisheries products is concerned. For example in 1989, most of the firms were established in Spain (41% of the total number of 963 firms), the U.K. 11% and in France 8.9%. The firms with the highest turnover in the Community are in the U.K.; 34.3% of a total of 7.5 million ECUs in the E.C., Germany 20.3% and Denmark 12.6% [126]. In terms of employment in the whole fish processing sector, the greatest work in U.K. firms, which employ 22.6% of the total of workforce 89,067 in this industry. Spain has 18.6% and Germany 15.5% [148,174]. Thus, there is a considerable structural heterogeneity in the Community's processing industry, and the mix between family business and multinational processing firms varies considerably between the Member States. In addition, there are no data available on a consistent basis for all Member States that can show the reliance of this sector on the Community's fish supply, and consequently the relation of the Community's fishing sector to the processing industry is once more left to intelligent guesswork. This is underlined by the fact that this type of the industry relies on the seasonal supply of fish and as a consequence its canning operations can switch to other food products or labour can be employed seasonally. For these reasons any absolute statistics about the canning industry must be treated sceptically (e.g. DG XIV own commissioned studies and Commission's Regional Socio-economic Studies). However, the total commitment appropriations of the Community for the processing industry during the years 1983 to 1993 represented 16% of the Community's expenditure on structural programmes.

The term auxiliary industries, (to the fishing industry) is used to describe all these activities that are directly related to the fishing industry. For example, the construction and repairing of fishing vessels and gear; various types of technical services, the maintenance and the facilities of fishing ports. Unfortunately, there are no robust data to indicate the importance of auxiliary industries to the Community's economy. There are some crude estimations that every job at sea generates five others on land and that the number of workers in related industries is estimated at around 2 million people [167]. Whatever the actual number of the workforce it goes without question that to integrate such a multi-dimensional industry under a management scheme needs more effort for the better understanding of all parameters. A systematic and uniform monitoring system, which would go along with more bioeconomic and socioeconomic analyses, would be of indisputable value.

At the moment such an approach represents only the fourth and fifth priorities out of the five major Community approaches in the C.F.P. as they appear in the Community's expenditure (figures 5.1 and 5.2). This amount of expenditure is very modest for the better understanding of an area that is claimed to represent one of the most integrated policies of the Community. In addition to the complex nature of the fishing industry, the Community's legislation for fisheries burdens the system with additional complications.

Resource management: Access restrictions

The conservation policy of the Community is based on a variety of technical measures adopted on the basis of scientific advice and covers the following areas:

1   regulation of mesh size in fishing nets;
2   regulation of fish sizes that may be retained on board or landed or offered for sale;
3   regulation of fishing gear and appliances (other than mesh sizes);
4   the establishment of open or closed seasons;
5   the establishment of open or closed areas;
6   restricted access to certain areas for licensed vessels according to their size and engine capacity;

7   regulation of time spent at sea for certain vessels.

The bulk of the technical conservation measures is contained in a 1986 regulation [40] that controls the mesh size of nets, the size or minimum weight of fish landed, the power of vessels, closed fishing seasons and the fishing gear used. These measures are applied individually or in combination to certain vessels. Other conservation measures for protecting the biologically sensitive areas are the establishment of the "Shetland Box" [31] (introduced in 4.5.1) and the "Norway Pout Box".

The "Norway Pout Box"

The "Norway Pout Box" originated in 1977 as an autonomous measure by the UK and was adopted in part by an E.C. conservation regulation in 1977 [41]. According to the British position, the aim of this enclosed area of the North Sea off North East Scotland was to protect juvenile stocks of haddock and whiting from the industrial fishing, mainly by the Danes, for Norway pout. The measure permitted a certain percentage of by-catches (from 5% to 10%) of herring, when fishing for sprat or Norway pout, and also banned the carrying of nets of various sizes on the same voyage. In mid-1978 the UK decided unilaterally to extend (2° eastwards) the box from 1 October to 31 March each year. In the Council, Denmark refused to accept the unilateral ban imposed by the UK. In addition, two judgements of the European Court in 1979 and in 1980 [42] ruled against the UK extension of the "Winter Box" based on the belief that it was not purely a conservation measure because of its unilateral character introduced by the UK and not agreed by the other Member States. Finally, the dispute was settled by the Council agreement in September 1980 as part of conservation measures involving restrictions on mesh sizes, types of gear, by-catches, closed seasons and areas [33]. This compromise produced a "Norway Pout Box", smaller in size, where the Danes were allowed to catch Norway Pout in the previously extended "Winter Box".

Access control after the Portuguese and Spanish accession

Further measures that restrict access to the waters (other than Mediterranean waters) of the "ten" were imposed on the new Member States, Spain and Portugal, under the articles 157 to 160 of the Act of Accession in 1985 [43] and the E.C. regulation 3531/85 [44]. The restriction of access arrangements were adopted for a period of 17 years from 1 January 1986 to 31 December 2002. First, articles 158 and 159 regulate the fishing possibilities of 450 Spanish vessels for non-specialised fishing in ICES divisions vb, vi, vii, viiia and viiib. They also ban (articles 158 and 349) Spanish and Portuguese vessels from fishing in a previously defined area around Ireland ("Irish Box") until 31 December 1995 when possible adjustments may be made. For the same period 170 vessels, on average, of the "ten" (articles 164 and 351) are permitted to fish for albacore tuna and tropical tuna in Spanish waters. As for Portuguese waters, according to article 351, the fishing activities of the "ten" are restricted only to pelagic species and to species not subject to TACs and quotas in ICES divisions ix, x and in the Fisheries Committee for Eastern Central Atlantic (CECAF) area. Fishing for albacore tuna and for tropical tuna is only permitted for certain types of vessels and fishing methods and for certain periods of time and areas. The fishing possibilities and the corresponding number of vessels of all sides have been amended several times, since the Act of Accession, almost annually from 1986 [45]. Since then, fishing effort is restricted by means of the so called "basic" and "periodic" lists. For example, since the 1985 "basic" list of 300 Spanish non-specialised vessels, an average of 285 have been permitted to fish in the 1989-92 period. According to the "periodic" list of 1985, 150 non-specialised vessels have been permitted to fish in the waters of the "ten", and in the "periodic" lists of 1989-92, 144 vessels have been authorised [46]. All the fishing vessels authorised by the lists to fish in certain regions, depending on their nationality, are grouped into geographical categories and meet certain restrictions that concern their technical characteristics, fishing zones, species they fish, the period of authorisation, the fishing gear they use or hold on board and the minimum consecutive number of days at sea. The vessels in these lists that are specialised in a particular type of fishery have an obligation to carry a licence document on board, issued by the E.C., that contains information about the identification of vessels, their owners, the period of the fishing authorisation and the type and the geographical zone of fishing, where applicable. Furthermore, under the Act of Accession, Spanish vessels, other than tuna type vessels are not permitted to fish in the triangle that is constituted by the French waters in the North-Eastern part of ICES division VIIIc. However, the French authorities face considerable difficulty in controlling Spanish vessels in the whole of area VIII, because of false or masked identification used by Spanish vessels, although this type of activity is prohibited by an E.C. regulation [47].

The above measures aim at the conservation of the resource by controlling the "input", by restricting vessels and access, and balancing fishing effort with available stocks. Although the above measures, that regulate access between the two new Member States and the other "ten", have been carefully designed to serve this aim, by applying licences and by means of "basic" and "periodical" lists, in practice they have proved to be less effective than it was first planned.

This happened for two reasons. 1) Either the TACs are set in excess of the existing fish population and therefore there are not enough fish in the sea to catch, or 2) it is not economically efficient to catch these TACs. The following data (table 5.1) support this remark. They show the average level of utilisation of fishing possibilities for fisheries, to which access is restricted by vessels in the lists and those that were licensed to fish for the years 1989-1991. It can been seen that the level of utilisation varies quite considerably from 46% for 110 vessels of the "ten" that fish for albacore tuna in Portuguese waters to 96% for 435 Spanish vessels permitted to fish these type of fisheries at that period (data published in the SEC(92) 2340 final Annex II, III [46]). There is pressure for regular amendments of the lists by Member States and these show a decreasing trend in the number of vessels permitted to fish each year. However, the solution is not a simple matter of decreasing permitted vessels. A more effective measure would be to establish a unified method for estimating the fishing capacity of vessels (engine power, fishing gear) and to combine different means of controlling exploitation levels, for example authorised amount of catches, specific fishing zones and licences. Such proposals will be discussed in the final chapter.

More effort should be devoted to understanding the special characteristics of each fishery before the Community produces the lists and donates the right to fish to Member States. For example, the utilisation of fishing possibilities for the specialised fisheries varies from 0% to 100% for all the parties concerned. Specifically, Portuguese vessels make almost no use of their fishing opportunities in the waters of the "ten". Another case that shows limited consideration of several parameters of a problem was the difficulties that Spain and France experienced in exploiting their quotas for monkfish in ICES division VII and for anchovy in ICES division VIIIc; this led to an interchange of their quotas in 1992.

"Quota Hopping" practice and critique

A severe dispute, mainly between the UK and Spain, arose over differences in the interpretation of the principle of equal access. The principle of equal access is well established in article 2(1) of the E.C. regulation 101/76 [48] and states that the regulations laid down by each Member State governing fishing in its maritime waters must not be discriminatory and distinguish between different fishing vessels or fishermen on the basis of their nationality.

117

Table 5.1 Average level of utilisation of fishing possibilities as restricted by the number of vessels in the lists and those licensed according to the arrangements in the Act of Accession of Spain and Portugal

FLEET CONCERNED	1989-91
	% of utilisation
Spanish "non-specialised fleet" in the EC-10 waters	96
Spanish "specialised fleet" in the EC-10 waters	82
Spanish fleet in Portuguese waters	76
Portuguese fleet in the EC-10 waters	0
Portuguese fleet in Spanish waters	84
EC-10 fleet in Spanish waters	40
EC-10 fleet in Portuguese waters	46

source: SEC(92) 2340 final, 23.12.1992

118

The "quota hopping" practice involves the transfer of fishing vessels from one Member State to the flag of another Member State in order to use the latter's fishing quotas. With the Spanish and Portuguese accession in 1986, the UK felt threatened that Spanish vessels would fish against UK quotas and laid down new licensing conditions for all vessels that were fishing for UK quotas and were over 10 metres long. In 1988, the UK introduced a new Merchant Shipping Act that aimed to prevent mainly Spanish vessels, largely owned by the Spanish-backed company, Factortame Ltd., from deriving economic benefit by exploiting UK quotas and to ensure that the already licensed vessels had a real economic link with the UK. Spanish vessels tend to fish off the West coast of the UK, mostly off the South-West of England and Northern Ireland, and they concentrate on shellfish and on species which are not targeted by the UK fleet, such as hake. Under this Act of 1988 a vessel, in order to be eligible for a licence, ought i) to be registered as a British fishing vessel ii) to operate normally from UK, Channel Islands or Isle of Man ports by landing and by regularly visiting these ports and iii) to have at least 75% of its crew from British or E.C. nationals ordinarily resident in the UK, Channel Islands or Isle of Man until 1992, when the transitional period relating to the free movement of all E.C. workers stops. Spanish workers were not considered as E.C. nationals during this period. iv) All crew members must contribute to a national insurance scheme in the UK, or the Channel Islands or the Isle of Man.

It is not clear what the actual number of "flag-of-convenience" vessels directly affected by this act is. Different sources estimate from 150 vessels that originally de-registered to as few as 16 boats. The Fisheries Minister, Mr. David Curry, in 1990 claimed that only 53 vessels were actually covered by the act, 37 of them already fished in UK waters and, therefore, 20 vessels were directly affected. The Sea Fish Industry Authority, in a paper on quota-hopping sent to the House of Lords Select committee on the European Communities [49], asserts that almost all these vessels were Spanish except for 6 Dutch beam trawlers fishing from Scottish North Sea ports, a Norwegian fishing from Shetland and another Dutch vessel fishing from the English port of Lowestoft. The most reliable data on the quota hopping practice come from the Scottish Office, Agriculture and Fisheries Department and are presented in table 5.2.

In December 1989 the European Court of Justice in an interim judgement [50] ruled against the 1988 UK Merchant Shipping Act, stated that Member States may not impose restrictive conditions, relating to the nationality or place of the residence of the vessel's crew and permitted the option of landing catches in a country other than the flag state. They also declared the vessels freedom to be engaged in "normal" fishing activities. The Court, additionally, suspended the application of the Act pending the outcome of the full Court case. As a result of

Table 5.2 Vessel balance sheet related to the "quota hopping" practice

Figures given by the Scottish Office, Agriculture and Fisheries Department

Vessels de-registered on 31.3.89	95 Factortame vessels 55 non-Factortame vessels } 150
Vessels re-register between 1.4.89 and 3.7.90, either following their sale to qualified owners or -in one case only- as a result of the European Court's interim relief ruling in respect of the nationality requirement.	20 Factortame vessels 12 non-Factortame vessels } 32
Factortame vessels registered following the House of Lords interim relief in July 1990	40 out of the 95 de-registered, and 8 more, Factortame vessels 5 more non-Factortame vessels } 53
Remaining unregistered vessels	69
Remaining Factortame vessels (of which 13 are entitled to benefit from House of Lords interim relief)	25
Non-Factortame vessels remaining unregistered ("many" of them have sunk, been scrapped or re-flagged or are now engaged in non-fishing activities)	40

Source: House of Lords, select Committee on the EC, "Review of the Common Fisheries Policy", session 1992-93 (2nd report)

this ruling only one vessel re-registered and 31 other vessels out of the original 150 vessels were sold to qualified owners in order to keep in business (Table 5.2). After the Court's interim finding, the UK government, foreseeing that the whole act would be declared illegal by the Court, tried to defuse the matter by the "House of Lords interim relief ruling" in July 1990. This entitled 66 vessels of the original 150 de-registered vessels to re-register and 53 did so (Table 5.2).

In July 1991 the European Court of Justice [51] made legal history by ruling that the U.K.'s 1988 act was illegal and should be amended because it was "a clear violation of the right of the Community citizens to exercise an economic activity in any Member State". It was the first time that the E.C. overruled a Member State's act of Parliament by declaring that the dominant principle must be that of non-discrimination between nationalities and that of free establishment for companies. This judgement was based upon the articles 7 and 52 respectively of the Rome Treaty. The European Court also ruled that compensation to vessels affected by the UK's act must be met by the UK. According to a lawyer representing vessel-owners affected by the act, around 35 of them did not seek compensation "out of the goodness of their hearts". However, claims from other vessels are estimated to amount £20 millions.

At the moment 90 out of 95 Factortame vessels have re-registered and in addition about 25 new vessels have come from Holland to register in the UK. According to the Fisheries Secretary, Mr. R. Carden in January 1992 [49] the remaining 60 vessels of the 150 that can potentially come back, have to meet the requirements of registration in the UK, i.e. safety rules and others, and it is estimated that only half of them would be met. They also have to acquire a licence to start fishing. Licences are not issued any more and they are not given for free. They have to find someone who already has a licence and he is willing to part with it, or he wants to sell a licence.

As a critique of the "quota-hopping" practice it can be mentioned here that the conflict between the UK and Spain is not the only dispute of such a kind in the Community. The Commission, which took a position in support of the Spanish vessel-owners, has also the obligation to act against other Member States whose criteria for licensing fishing vessels discriminate to a similar extent on the grounds of nationality. For example, there are Dutch-owned vessels in Germany, Spanish vessels in Ireland and, ironically, Spain has adopted on the one hand a licensing system discriminating on grounds of nationality in her waters, while on the other hand is applying pressure on the E.C. for the replacement of national quotas by a single E.C. quota system.

The problem of "quota hopping" and the European Court judgement of July 1991 are not as straightforward as they seem and need more widespread attention. First, if vessels with flags of convenience increase, they will create more problems for the UK in meeting its obligations under the Multi-Annual Guidance Programmes (MAGPs), which will be discussed further below. Second, by using this ruling vessels from another Member State, that has a strict quota management regime, can utilise a flag of convenience from another Member State and thus return to their country's waters and fish without any national control. This is the case of the above mentioned 25 Dutch vessels, which in order to avoid mainly the restriction of not going to sea for 150 days per year, as the Dutch fishery management system requires, managed to register in the UK and thus fish without any control back in their country. Third, another problem is that the ruling of July 1991 testifies that the principle of freedom of establishment is superior to "relative stability", on which the whole structure of the quota system is based and which has been accepted by all the Member States. Therefore, as the E.C. moves towards a single market, the national quota system seems not only to lose its effectiveness as a conservation measure but also becomes somewhat incompatible with Community Law. However, the judgement did not change the principle of relative stability, and furthermore the 1992 Council regulation [155] for the period 1992-2002 is to retain the TAC system and relative stability with minor adjustments to reflect existing fishing patterns. For such arguments it may be possible to blame important lacunas in the E.C. legislation and the lack of foresight about more or less obvious possible conflicts and questions. Often it has become common practice to deal with problems after they have become acute. However, dealing, resolving and proposing new measures are not easy tasks in the E.C.. Not only the bureaucratic structure between the institutions and various committees in the E.C. but also the procedure of trying to achieve a compromise has to be blamed.

Technical conservation measures

The categories of technical measures mentioned above, covered in the 1986 basic technical regulation [40], are regarded as "permanent". However, this regulation has been amended several times; there have been twelve revisions with the eleventh and twelfth revisions pending before the Council. The controversial aspects of these technical measures are discussed below. These concern mesh sizes and tie-up measures.

On 28 October 1991 the Council adopted certain technical conservation measures, amended in 18 February 1992 [55], to increase mesh size in certain regions. For example, in the ICES areas IV (North Sea) and VI (West of

Scotland) the regulation increased the minimum mesh size for fishing demersal fisheries from 90mm (diamond) to 100mm (diamond) with the option of 90mm of square mesh size. Furthermore, the Community brought in a tie-up scheme for vessels operating in the above areas and fishing for demersal species. This scheme was for vessels to be tied up for eight consecutive days per month. These measures came into effect on 1 June 1992.

According to European Court judgements mentioned above and specifically according to the case 53/86 [53], Member States can take national technical measures imposed only on that nation's fishermen, that are stricter than the Community measures. The U.K. suggested a mandatory requirement, applied from 1 July 1991, to fit square mesh panels of 90mm to vessels of 400 horse power instead of the 90mm diamond mesh size. However, this unilateral measure did not find great enthusiasm in the Council, which is determined to return to the issue of mesh size no later than 1 January 1995, if scientific advice confirms that the current measures have failed to achieve their conservation purpose, which is a 30% reduction in catches of all age groups or a sufficient reduction in the smaller age groups for demersal fisheries.

The eight days tie-up measure was universally condemned for being "anti-conservation, dangerous and anti-social" and therefore on 17 December 1991 [56] the Council produced an acceptable compromise of 135 days tie-up from the 1 February 1992, with complete flexibility for fishermen to choose the days suitable for them. According to the UK Association of Fish Producers' Organisation and specifically according to the chief Executive of the Scottish Fishermens' Organisation, Mr. MacSween, this tie-up measure does not burden most vessels, as "very few boats fish for 230 days in any case". As the "Fishing News" in 1991 [52] noted, the two beam trawlers with the highest earnings fished only for 260 and 254 days in the absence of the eight days tie-up rule. Nevertheless, indicative of how controversial are such rules, the Scottish Fishermen's Federation requested from the UK government financial compensation for vessels affected by the rule.

For 1993, the press reports that another more sophisticated scheme will be proposed by the Community, which will allow fishermen to use nets bigger than 110mm diamond mesh size up to 120mm for fishing roundfish (cod, haddock, whiting, saithe etc.). With the increase from 110mm to 120mm, fishermen would benefit from a decrease in tie-up days from 8 to no tie-up respectively.

It is too soon to assess the effectiveness of the square mesh size net in conserving fish stocks by allowing small fish to escape, by reducing discarding and retaining the marketable size of fish. However, there is now much evidence gained by commercial seagoing trials from the Scottish and the Northern

Ireland fishing fleet. The results of these trials were presented by the Scottish Fishermen's Federation [49]. During 1990, vessels took part in these trials by fitting into the conventional 90mm diamond shaped mesh size net, a window of 80mm square shaped mesh size net. The result was that a 31% reduction in haddock discards and a 40% reduction in whiting discards while the meshes of the cod end were open irrespective of the rigging of the trawl. With the same experimental design but fishing with a prawn trawl the Northern Ireland Fishermen's Federation declared [49] that the gear was not stable, was twisted, pulled and worked round. As a conclusion the square mesh net is preferable for fishing roundfish rather than flatfish. The Community's decision to permit square mesh nets would appear to be justified and it is almost certain to reach the objective of balancing the need to reduce discarding and retain marketable fish.

TACs and quota management system

The TACs and quota management system are the corner stone of all the conservation measures taken by the Community. How this system works and to which areas and species it is applied has been described in 4.5.2. In the present section its importance as a conservation measure and its impact in the fishing industry will be discussed.

At present the Council fixes the TACs for 103 fish stocks each year of which only 39 are estimated with "sufficient" data. These are called "analytical TACs". The other 64 are called "precautionary TACs" and are based on intelligent and educated guesswork rather than on solid scientific data [167]. However, it has to be mentioned here that in the mid-term report produced by the Commission [54] it is suggested that TACs and the quota system must be retained with a few possible amendments for improvement. These could be to reduce the number of the precautionary TACs, to incorporate multi-species stock management i.e. to set TACs for a group of species instead of each single species, and to allow for pluriannual TAC determination i.e. to set TACs for several years at a time instead of one year as applies at the moment. The multi-species and multi-annual TACs (MSTACs and MATACs) have received extensive treatment in this mid-term report as they are supposed to be the panacea for all the deficiencies of the present system of TACs. However, in the new regulation of 1992 for the restructuring of the C.F.P. such new measures are proposed without being given any particular importance and without any detail and definition about their future implementation [155] .

124

The underlying uncertainty of the validity of fish population estimates was discussed in the theory of population dynamics. In order to explore the significance of the TAC system as a management tool and its effect on fishermen's practice, the case of the North Sea demersal fishery will be considered.

The demersal stocks in the North Sea are selected as the most representative stocks for assessing the TAC management system since their analytical TACs are among the best scientifically estimated in the Community waters. They are based on consistent analytical assessments of catch-at-age data using commercial catch and effort data and research vessel survey data. In addition, these stocks are selected because they are economically significant for a number of fleets in several Member States (e.g. The U.K., France, The Netherlands, Belgium etc.). They also represent more than three quarters of the catch of demersal species covered by TACs taken in Community waters during the 1981-1991 period [54]. The demersal stocks in the North Sea that are covered by TACs include six species, namely cod, haddock, sole, plaice, whiting and saithe. For each of these species, the scientifically recommended TAC, the TAC agreed by the Council, the actual catch and the spawning stock biomass (SSB) are plotted for the years 1981 to 1992, using data presented by the ICES Advisory Committee on Fishery Management (ACFM) [184] (Figure 5.4).

In order to explore fishermen's response to TACs and with which parameter the catches are associated better, the degree of association (or the closeness to fit) between catches on one hand and the recommended TACs, agreed TACs and SSB on the other are examined. The strength of the relationship between these variables is estimated by the sample correlation coefficient (r) (Table 5.3). Then, the correlation coefficient is tested for statistical significance at the 5% level. This was examined by employing the formula

$$t_{(n-2)} = \frac{r\sqrt{n-2}}{\sqrt{1-r^2}}$$

for the t-distributed random variable with n-2 degrees of freedom.

The maximum scientifically recommended level of TACs is used for the analysis. For sole, plaice, cod and saithe the actual landings almost equal the actual catches, which are used by the working groups of ICES and form a crucial basis for the estimation of SSB and consequently for the biological advice on TACs. However, for haddock and whiting there is a discrepancy between the quantity of landings and actual catch. This is because large quantities of these species are taken as by-catches mainly in industrial fishing

125

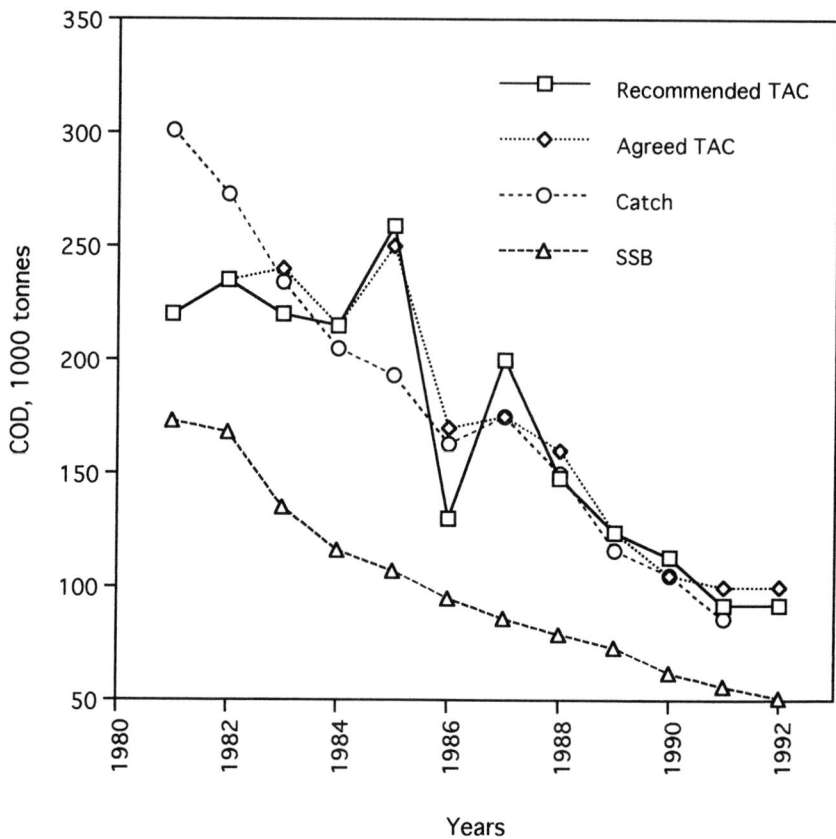

(a)

Figure 5.4   Evolution of the North Sea demersal
              fisheries covered by TACs; (a) cod,
              (b) haddock,(c) saithe, (d) whiting,
              (e) plaice and (f) sole

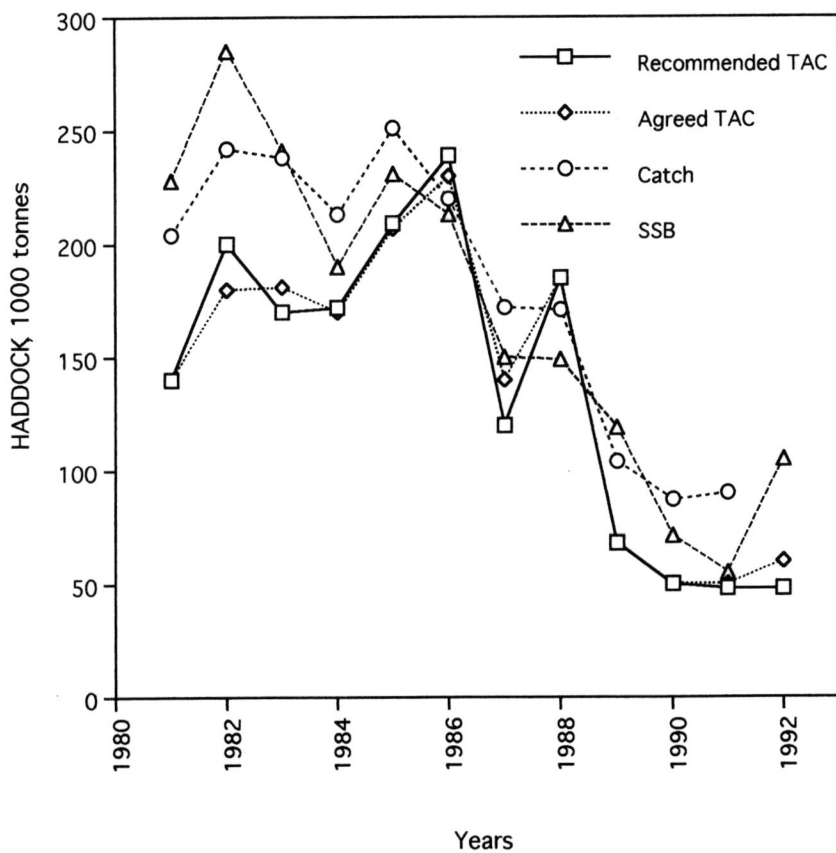

(b)

127

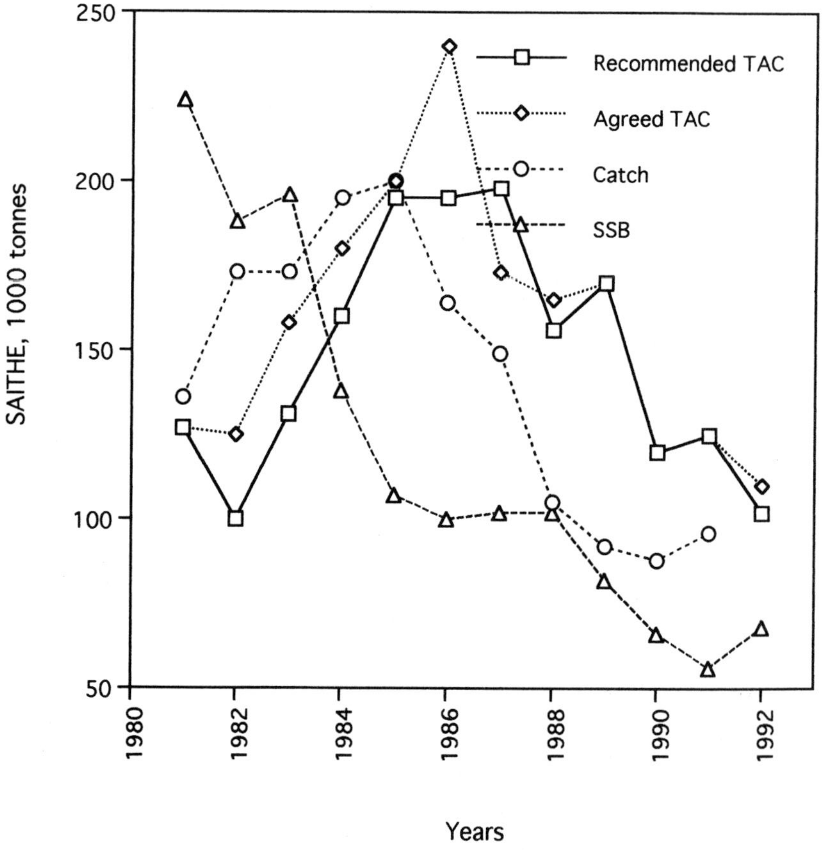

250
200
150
100
50
SAITHE, 1000 tonnes
1980
1982
1984
1986
1988
1990
1992
Years
Recommended TAC
Agreed TAC
Catch
SSB

(c)

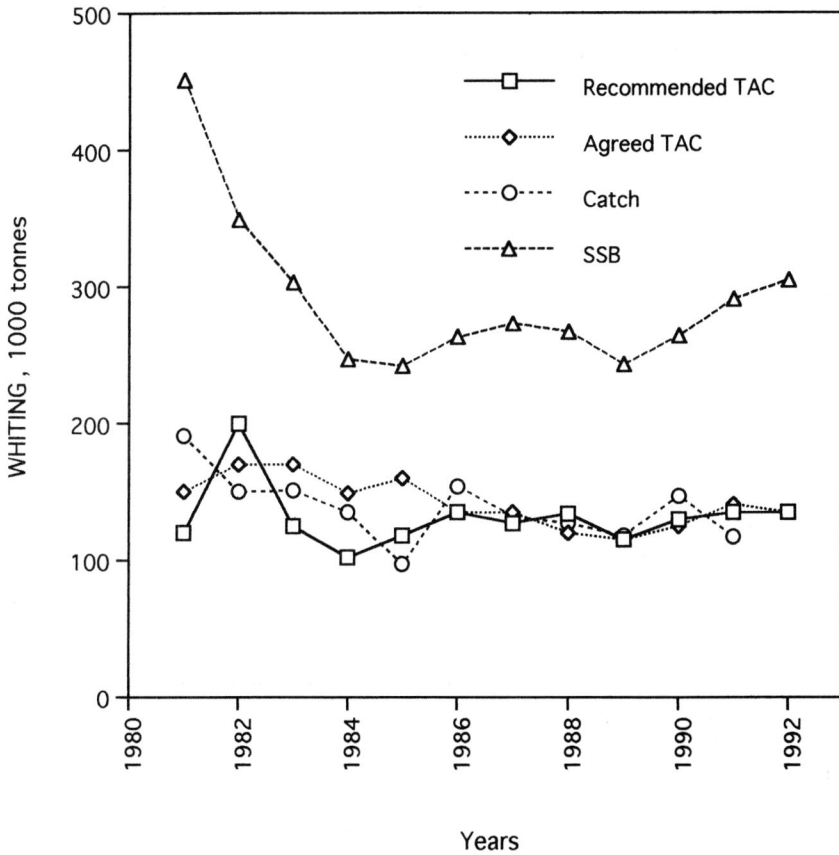

(d)

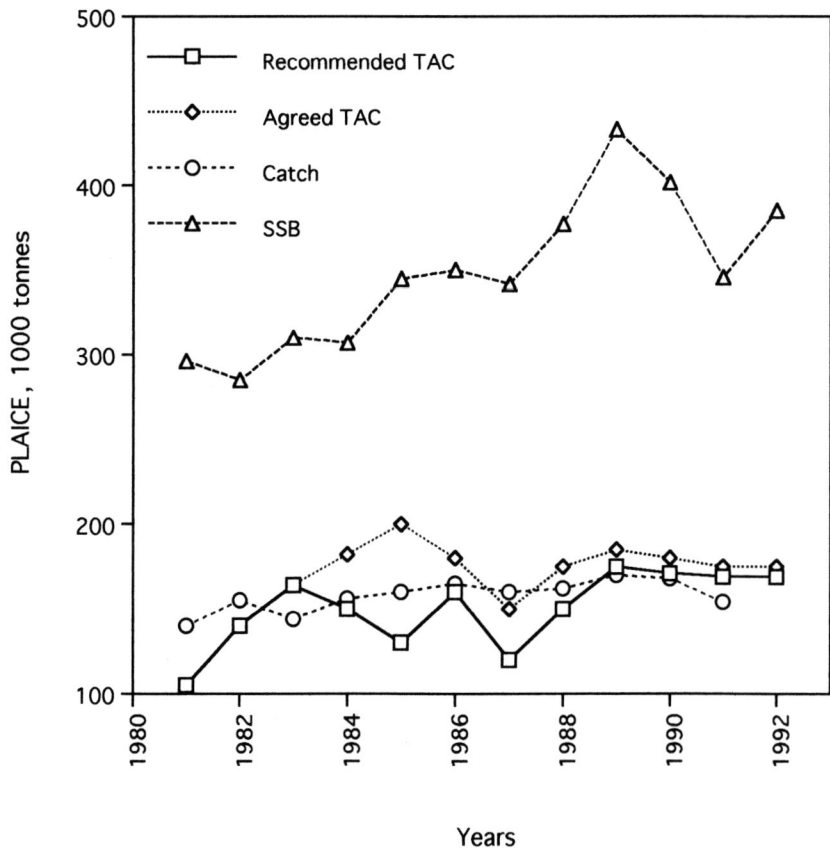

(e)

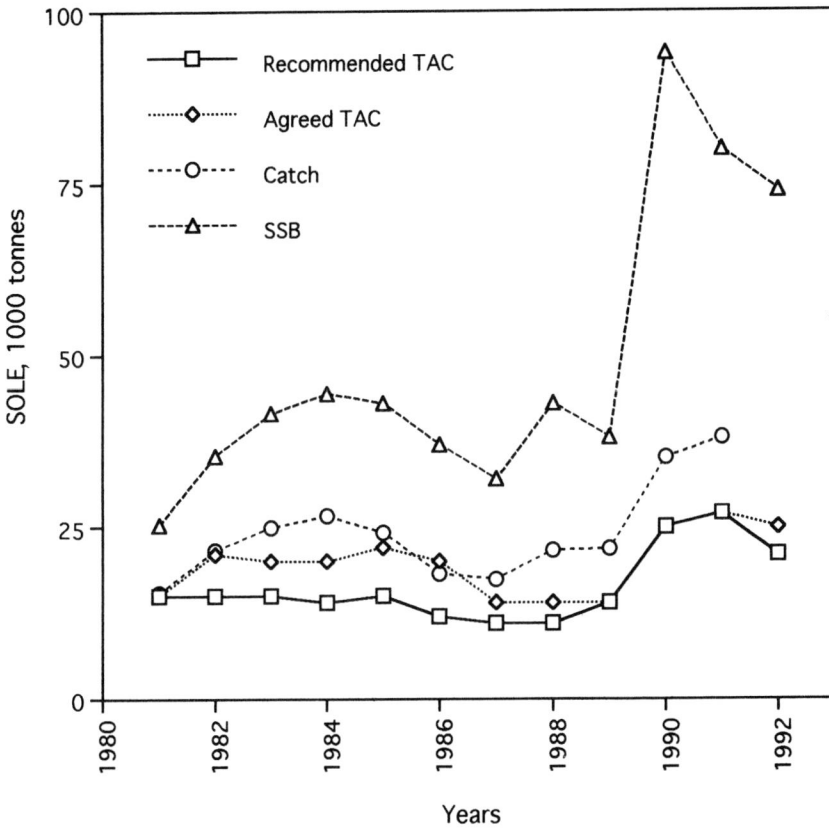

(f)

sources: ICES Cooperative research reports. Reports of the ICES Advisory Committee on Fisheries Management. Volumes: No193, part 1, chapter 3.5, February 1993; No179, part 1, chapter 3.5, February 1992; No161, part 1, chapter 3.5, February 1989

Table 5.3 Correlation matrices for North Sea demersal fisheries and significance test (t-test) for the correlation coefficients (r).

COD	Recomended TAC	Agreed TAC	SSB
Catch, (r)	0.8324	0.8671	0.9845
significance test (t)	4.5062	5.2221	16.8401

HADDOCK	Recomended TAC	Agreed TAC	SSB
Catch	0.9082	0.9148	0.9511
Significance test (t)	6.5098	6.7946	9.2375

WHITING	Recomended TAC	Agreed TAC	SSB
Catch	0.1726	0.2007	0.7915
Significance test (t)	0.5257	0.6146	3.8852

PLAICE	Recomended TAC	Agreed TAC	SSB
Catch	0.4951	0.6832	0.8048
Significance test (t)	1.7095	2.8068	4.0678

SOLE	Recomended TAC	Agreed TAC	SSB
Catch	0.8912	0.8216	0.9330
Significance test (t)	5.8940	4.3237	7.7777

SAITHE	Recomended TAC	Agreed TAC	SSB
Catch	0.2779	0.4723	0.5082
Significance test (t)	0.8679	1.6075	1.7702

and there is extensive discarding, which is not seriously estimated. For example, in 1991 official landings of haddock were 44,000 tonnes while discarding was estimated, by logbooks and scientific surveyors on commercial vessels, to be about 40,000 tonnes [185]. Similarly, for whiting in 1991 the officially reported landings were 46,000 tonnes, while discards added up to about 33,000 tonnes [185]. In figures 5.4(b) and 5.4(d) the actual catch of these two species which is used by the ICES working groups, is considered for the period 1981 to 1991, since it represents a more realistic approach to the quantity of haddock and whiting caught.

Cod

In general terms the recommended TACs appear to be reasonably close to agreed TACs for the period concerned (Figure 5.4(a)). The first divergence (20,000 tonnes) occurred when TACs were first legally established by regulation 171/83 [27] in 1983. Since then the oscillations in scientific advice were followed by the agreed TACs with the exception of 1986, where the Community's TAC was 40,000 tonnes more than the recommended level. During the decade 1983-1992 the TACs followed a downward sloping pattern since they were reduced from the 1983 level of 240,000 tonnes to the 1992 level of 60,000 tonnes. The response of fishermen to the TACs for cod seems to be to accept the agreed TAC, since catches above the recommended TACs in 1981 and 1982 have been succeeded by catches more in agreement with E.C. TACs since 1983. This is also shown by the strong degree of association (correlation coefficient $r>0.83$) between catches and all the other examined parameters (Table 5.3). However, catches are more strongly associated with the SSB ($r=0.9845$) rather than TACs. The t-test shows a highly significant correlation at the 5% level of significance (Table 5.3).

On the other hand, the state of the North Sea cod stocks is alarming. The decline of the SSB, which started in 1970 [185] continued and reached a record low in 1992. The downsloping pattern of catches seems to follow the more stable pattern of the SSB rather than the oscillating recommended TACs. Therefore the remark made by DeWilde 1991 that "even the scientific advice did not conserve the (North Sea cod) stocks" seems to hold.

Haddock

For haddock the recommended and the agreed TACs have kept more closely together than for cod (Figure 5.4(b)). However during the decade 1983-1992, TACs descended steeply. In 1983 the TAC was 181,000 tonnes but only

60,000 tonnes in 1992. Despite the TACs fishermen did not adjust their catches to the TAC level. This is also shown in the poor degree of association between catches and recommended, agreed TACs (t<6.79) (Table 5.3). However, there is a strong degree of association between catches and SSB (r=0.9511) (Table 5.3). The t-test shows also a highly significant correlation between catches and SSB at the 5% level of significance (Table 5.3). Haddock has been overfished most of the time since 1981. Although the North Sea haddock stocks have declined alarmingly since 1981 and reached a record low level in 1991, a recovery appeared in 1992, which, however, according to ASCF [185] is not expected to last long.

Saithe

The agreed TACs were variably higher than the recommended TACs during 1981 to 1986, with the greatest difference in 1986 of 45,000 tonnes (Figure 5.4(c)). Since that time both TACs have declined steeply. The Community's agreed TAC has been reduced by 54%, that is from 240,000 tonnes in 1986 to 110,000, in 1992. Catches have been declining more abruptly since 1985 possibly the result of overfishing the stocks during 1981 to 1985. There is a poor correlation between catches and all the other parameters examined (Table 5.3). The state of the North Sea saithe is very disturbing. The SSB was in 1991 at a historically low level but showed a slight recovery in 1992.

Whiting

During the period 1981 to 1986 there was always a difference between the agreed and the recommended TAC (Figure 5.4(d)). The agreed was higher in 1980 and from 1983 to 1985 with a relatively stable difference of about 45,000 tonnes. Since 1986 the Council has followed the scientific advice for TACs. There is a poor correlation between catches and recommended or agreed TACs (r<0.2). However, there is a strong degree of association between catches and SSB (r=0.7915) (Table 5.3). The t-test also shows a highly significant correlation at the 5% level of significance between these two parameters (Table 5.3). The state of North Sea whiting seems satisfactory since the SSB is at its highest level since 1983.

Plaice

As with cod the difference between the recommended and the agreed TACs started after the legal enforcement of TACs in 1983 (Figure 5.4(e)). Since 1984

134

the agreed TACs have exceeded the recommended. The highest difference occurred in 1985, when the Council instead of following the scientific advice for a reduction of 20,000 tonnes compared with the previous year, increased the TAC level by 70,000 tonnes above the recommended TAC for 1985. Catches, contrary to the oscillations in the recommended and agreed TACs, remained relatively stable during 1981 to 1991. There is a poor correlation between catches and recommended or agreed TACs ($r<0.6$). However there is a strong degree of association between catches and SSB ($r=0.8048$) (Table 5.3). In addition, the t-test shows a highly significant correlation between catches and SSB at the 5% level of significance (Table 5.3). The state of the North Sea plaice is satisfactory since the SSB is at a level above the long term mean.

Sole

The agreed level of TACs remained above the recommended between 1982 and 1988 (Figure 5.4(f)). Despite this difference catches were above the level of TACs since 1983 with the exception of 1986. There is a stronger degree of association between catches and SSB ($r=0.933$) rather than recommended ($r=0.8912$) or agreed ($r=0.8216$) TACs (Table 5.3). The t-test shows a highly significant correlation at the 5% level of significance for all the parameters concerned (Table 5.3). Despite the apparent overfishing of North Sea sole the SSB indicates that the stocks are in good shape. However, the increased SSB is likely to be misleading, according to the ACFM because of uncertainties in stock assessment [185].

In the case of sole and plaice the consistent overshooting of the scientific advice by the agreed TACs seems to have no apparent effect on the state of the stocks. Indeed these two species are in better shape than others such as haddock and cod where the scientific advice was more closely kept. Considering that the above stocks have been examined by stock assessment methods which are believed to be the most satisfactory the effectiveness of the TAC system for conserving these stocks is further questionable.

The principal conclusion of the above analysis is that the TAC system has not attained its objective to maintain the resource for the species concerned. For all the above species catches are more associated with the SSB rather than the agreed or recommended TACs. Overall, the TACs fail to represent the current catch situation which is more affected by the condition of the stocks in nature rather than by the imposition of TACs. This situation is indicative of the discrepancies in the conservation policy which fails to incorporate the economic and biological elements that define the fishing activity.

Fishery products have been divided into two main categories, i.e. species for human consumption and industrial species. Their relative contribution to total E.C. landings is presented in Figure 5.5. It can be seen that the proportion of landings attributed to industrial fishing varies considerably between countries but generally industrial fishing amounts for a quite small proportion of total landings with the exception of Denmark. Although the value of industrial species is considerably lower than the remaining fishery products, (for example in Denmark the price per tonne of industrial landings represents a mere 8.3% of that for human consumption) industrial fishing exhibits several interesting peculiarities which can affect more valuable species and which are worth being mentioned here.

This type of fishery uses fish stocks which cannot be consumed directly by man. The use of "industrial species" is to produce fish meal that is high in proteins and can be used for feed in aqua culture or in animal feeds, oils and fats and even to produce fuel-oil for power stations. Only recently (1990) Denmark decided to phase out gradually the use of fish oil to fuel power stations.

There are six major species that are utilised for industrial fishing that can be sorted into the following categories according to their human consumption compatibility:

1   Species that are never used for human consumption. Such are the Norway Pout, the sandeel and the capelin (which does not occur in Community waters).

2   Species that can also be used for human consumption but their primary use is for reduction to fish meal and oil. These are the blue whiting, the greater silver smelt, the horse mackerel and the sprat. Attempts have been made by the fish processing industry, specifically by the Germans and the Scots, to promote these species, mainly the blue whiting and the horse mackerel, for human consumption. Although there are huge quantities of them in the deep water west of the Hebrides and they are not difficult to catch, they have never found a place in the market for human consumption. This is primarily because they have very soft flesh and the processing industry cannot fillet them industrially.

3   Species whose primary market is for human consumption but in case of surplus catches they are sold for reduction to fish meal and oil. These are the herring and the mackerel. After 1977 when the North Sea herring

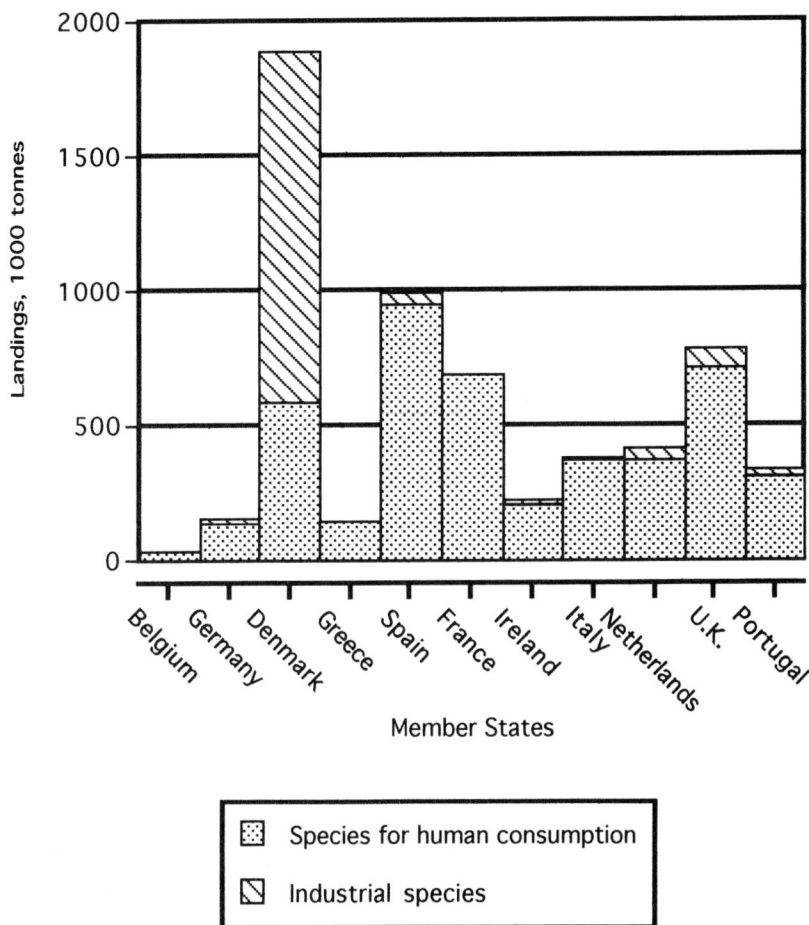

Figure 5.5 Total landings of species for human
consumption and industrial species.
Average quantities for years 1988-1989
(x1000tons)

sources: Commission of the E.C., Eurostat Statistical Office, "Fisheries: Yearly Statistics 1991", Luxembourg Office for Official Publications of the E.C., 1992 and D.G. XIV, XIV-A-3, 30.6.1993

stock collapsed a Council regulation (Regulation No 2155/77) prohibited the fishing of herring for industrial purposes. However, since the re-opening of the North Sea herring fishery in 1983, herring can be fished industrially, when the market is flooded, and there is no market for herring for human consumption.

In figure 5.6, a pie-diagram has been drawn based on the most recently published reliable data on industrial species landings of the 12 Member States during the years 1988 and 1989 and it shows the percentages of industrial species landings by Member States. Thus, Denmark caught 85% (1,299,481 tonnes on average during the years 1988-89) of the total E.C. industrial species landings (1,475,031 tonnes on average during the years 1988-89), followed by the UK with 4% (67212 tonnes on average) and Spain with 3%.(42234 tonnes on average).

The main problems of industrial fishing is that of setting TACs, the by-catch problem and the environmental issue of the importance of the fish caught to the food chain for "edible" larger fish or for seabirds or sea mammals.

All the species exploited only for reduction to fishmeal and oil are short-lived species; they are taken into the fishery in their first year of life and their populations show large scale fluctuations. Therefore, for the reasons explained in the second chapter, it is impossible to apply the analytical procedures of the VPA needed to obtain estimates of recruitment in time in order to set analytical TACs. Thus, a large percentage is not covered by analytical TACs and the precautionary principle is applied. TACs are calculated without sufficient scientific support because the exact size of the stocks is unknown. They are applied in order to avoid large-scale inputs of effort for this type of stocks.

It has been mentioned before that the minimum mesh size for nets used to catch fish for human consumption varies usually from 85mm to 120mm. For industrial fisheries the net size used has minimum size of 17mm. This tactic has been much criticised by the UK, particularly by the Scots, and also by the French fishing industry, as it is incompatible with the priority given to conserving fish stocks. It is obvious that the use of such a small mesh size will produce large quantities of by-catches of juvenile and mature fish for human consumption. The permitted quantity of by-catches set by the Community must not exceed 10% of the weight of the total catch retained on board. However, it is very difficult, if not impossible, to identify the quantities of the various species caught as by-catches. This is because the landed catch is usually crushed by the gear and becomes a pulp of fish, which is often difficult to identify.

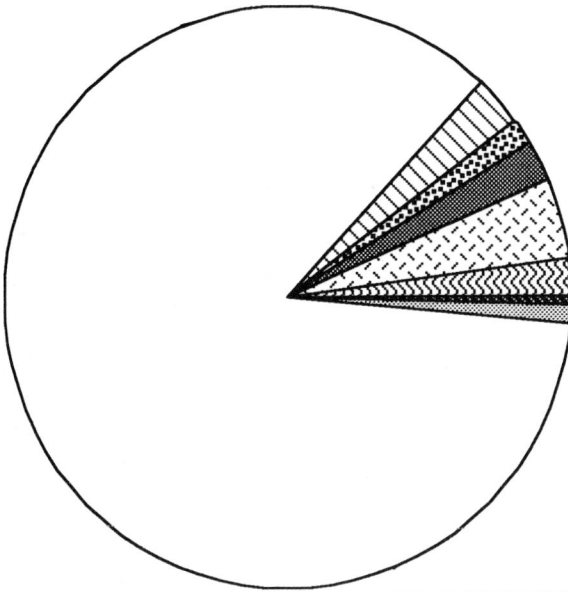

▨ Others (1%)	▦ Germany (1%)
☐ Denmark (85%)	◩ Spain (3%)
▦ Ireland (1%)	▨ Netherlands (2%)
▨ U.K. (4%)	▨ Portugal (2%)

Total E.C. average landings of industrial species in 1988 and 1989:
1,475,031 tonnes

Figure 5.6 Percentages of the landings of
industrial species by the E.C. Member
States during the years 1988-89

source: Commission of the E.C., Eurostat Statistical Office, "Fisheries: Yearly Statistics 1991", Luxembourg Office for Official Publications of the E.C., 1992

Haddock and whiting are caught as by-catch in the Norway pout fishery, while in the sprat fishery young herring is taken as by-catch. For the latter, because the sprat stock collapsed in 1989, for reasons attributed to environmental perturbations, almost the entire catch of "young clupeoids" (herring and sprat) consists only of herring. The Community reacted and adopted a regulation on July 1992 [186] by setting a minimum of 32mm mesh size to fish sprat in the Scagerrak and Kattegat and limiting the by-catch to a maximum of 5%. The industrial fishing for blue whiting, horse mackerel and for sand eels produces a relatively "clean" catch, free from any by-catch of the protected species.

The species which are taken in industrial fishing are small and tend to be low down in the food chain. For example, in the Baltic ecosystem there is one major predator, cod, and two species of prey, sprat and herring. In the North Sea, there is a close interaction between a large number of species, living in a relatively stable physical and chemical environment. The North-East Atlantic and the Barents Sea ecosystems are similar to that of the North Sea but with fewer predators, mainly cod and haddock, which prey on herring and capelin. Therefore, an assumed collapse of those prey species that are fished for industrial purposes may have detrimental effects on fish higher in the food chain and even on birds and mammals. For example, according to a report, prepared by M. Holden for the "Select Committee on the E.C." of the House of Lords, a total ban on the fishing of Norway pout would result in an increase in the average landing of haddock by 5,000 tons and whiting by 43,000 tons with a total value of £48 million [49]. On the other hand, Holden stresses that this is not a clear-cut analysis and the benefit from closing down the industrial fishing is ambiguous, because the discarded haddock and whiting caught for human consumption is estimated at about £90 millions. He concludes that "it seems inappropriate that the fishery for Norway Pout should be banned at least until such time as those who prosecute the human consumption fishery for haddock and whiting and who benefit from such a ban stop discarding". An example of the effect of industrial species depletion on bird populations is the interaction between sand eels and the Arctic Tern (*Sterna paradisaea*) (Monaghan *et al.*, 1989). When the sand eel stock around Shetland declined in 1980, it was noticed that the Scottish Arctic Tern population dropped dramatically. A comparative study by the Zoology Department of Glasgow University was carried out between the birds' colonies in Shetland and Coquet Island, where there is no industrial fishing. The study showed that the Shetland terns were unable to find sand-eels of intermediate size (4-8cm) to feed their chicks, which died. The study concluded that it is not possible to assert whether the lack of sand eels was due to industrial fishing or to environmental reasons (Monaghan *et al.*, 1989).

The various studies could not, given the present knowledge, include all the parameters and characteristics of the ecosystem that would enable us to conclude with certainty about the effect of industrial fishing on the fishing industry and the environment. In making recommendations for TACs, the ICES recently made efforts to estimate fish mortality levels, not only by taking into account natural and fishing mortality but also the mortality attributed to predation by birds and mammals. A theoretical bioecomonic model, which aims to incorporate environmental degradation in a price mechanism for a proposed fisheries management scheme, will be presented in the final chapter.

The efficiency of the TAC resource management policy

It has been mentioned before that, out of some 103 fish stocks that are managed under the TAC system only 39 are estimated with "sufficient" biological data and are called "analytical TACs". The TACs for the other 64 fish stocks are based on intelligent and educated guesswork rather than on solid scientific data and are called "precautionary TACs" [167]. The coverage of species by the TAC system is not distributed uniformly in the Community's waters. The coverage is higher in the north and diminishes in the south with the absence of any fish stock under the TAC system in the Mediterranean. It has been shown (Figure 6.1 page 150) that the unit value of landings is typically higher in the southern regions of the Community. Moreover, the value of the landings of fish that are covered by TACs (analytical and precautionary) represented only about 17% of the total value of Community landings (17,648 million ECUs) between 1987 and 1989 [54]. Scientists evaluate the state of fish stocks covered by TACs based on an arrangement which classifies them according to the effect of fishing mortality on the average yield condition. This indicator, though somewhat arbitrary and uncertain, shows that about 77% of the analytical TAC stocks are either fully exploited or depleted [54]. Therefore, taking into consideration also the structural development of the Community's fishing industry (see previous chapter), the TAC and quota system does not seem to have had the desired influence on the direction of economic development in the fishing industry. There are certainly some sectors of the fishing industry in certain Member States, mainly in the northern regions of the Community, where the TAC and quota system plays an important role in the industry. However, it has been estimated by the Commission that the overall contribution of non-quota species to the value of total landings in Member States that have traditionally depended on species subject to TACs ranges from 25% to 100%. (167).

Furthermore, under the existing TAC system there has been an overall gradual under-utilization of quotas over the years. For example, in 1983 when the TAC system was first introduced, unused quotas amounted to 5%, in 1988 it was 16%, in 1989 15.3% and 1990 saw the lowest ever utilization of quotas with 27% unused (167,54). There is a series of incidents resulting in unused quotas which may be detected from the setting of the TACs until the time they are utilized.

Scientists may overestimate quotas as they have to deal with the unpredictability of the fish population (as explained in the second chapter) coupled with the indeterminate nature of the precautionary TACs and the unreliability of the given data. This could result in TAC implementations which are impossible to be complied with, since there is simply not enough fish in the sea to catch. Additionally, knowing the Community decision-making procedure, it is almost the rule that under political pressure and compromises the scientific advice is often altered and revised upwards in the Council of Ministers, as happens in the North Sea demersal fisheries (5.4.1 and DeWilde, 1991). As a consequence there is an increase in "paper quotas" which as their name implies are to be found only in paper.

From the time that quotas are allocated to fishermen the possible scenaria for their under-utilisation are more complex. Maybe it is not simply and solely that there is an inadequate supply of fish in the sea. It could be that under the pressure of balancing the cost of the increased capital intensity of the fishing fleet (discussed in chapter 6), sectors of the fleet, which are most likely to be the larger vessels, have to diversify their effort to fishing for non-TAC species and high-valued species in order to maintain their viability. In addition, this diversification can be attributed to the advent of externalities (see page 9), when taking into consideration the reported nomadic behaviour of fishing vessels (Sampson, 1991). Unfortunately, it is not possible to establish the contribution of economic parameters to the under-utilization of TACs, since there are no data available.

On the other hand it is likely that the TAC and quota system along with the reduced trend in catches since the Spanish and Portuguese accession in 1986 (figure 6.4 page 160) have been an additional incentive for fishermen to expand the fishing activity in order to utilize the catch possibilities in an accelerated way. Therefore, they would try to take their share before TACs are taken by other fishermen. Unfortunately, although the overcapacity in the Community has been taken for granted and there are various analyses as to how much it has to be reduced (for example see (159)), there are no efficient and robust data to support such estimates. It is not only the uncertainty and lack of economic and social parameters or the theoretical deficiencies in population dynamics theory in attempting such an estimation, but also the unreliability and inefficiency of the biological data about fish stock abundance in Community waters that require any estimations of capacity reduction to be taken only as theoretical, indicative measures. For example, the small number of analytical TACs, the partial reliance of the fishing industry on fish stocks covered by quotas and the degree of fraud that the system generates are a few main elements that create uncertainty and "noise" in bioeconomic models of the Community's fisheries.

As the Commission's mid-term report admits [54] there is limited compliance with TACs and quotas everywhere in the Community. The Commission has estimated [54] that a significant discrepancy exists of up to 60% between the reported figures for catches and the real ones. In addition discarding at sea is considered to be a generalized phenomenon that concerns all types of fisheries. In the Bay of Biscay discards for hake were estimated to exceed the landings measured in individual fish [54]. However, data on discards are often unavailable and as a consequence it is not strange that a "black" market in fish has developed. Although its existence has been admitted by the U.K. National Federation of Fishermen's Organizations [161], its dimensions are swept under the carpet. As a consequence, only the official declared landings are to be debited from quotas, thus making the system more vulnerable to biases.

Discarding is not solely generated by the current TACs and quota system and the technical conservation measures that require a certain marketable fish length and a certain size of by-catches. Discarding in North Sea fisheries (mainly of haddock and whiting) was recorded long before the application of the TAC system in the 1960s. Discarding is also common in the Mediterannean where there is no TAC system. There are also economic incentives, which go along with the conservation system, that nourish this practice. For example, the TAC system "encourages" fishermen to retain the larger and more valuable fish in order to maximize the economic benefits from their quotas. Similarly, in the case where a catch consists of species with no commercial potential or when the market price of these species is lower than usual or there is no market potential in the country of the vessel fishing for the species, discarding and fishing for more valuable fish may be a rational economic solution when the costs of fishing are also considered. Furthermore, when the costs of fishing are increased by long trips or by storage and packing facilities on board or other reasons, the targeting of high value fish inevitably produces more by-catches not only of other species but also of the target species, when the fish are small or are damaged by the gear used. Finally, under this scheme of TACs fishermen are obliged to discard species for which their quotas have been exhausted, although the fish were caught in a mixed fisheries catch.

The future of the resource management

If the future of resource management were to evolve directly from the new regulation of 1992 [170] it would be compatible with the style of this regulation, oblique and obscure. Hopefully, adopting the most pessimistic view, after political bargaining and compromise, a better future should arise than that assumed in the regulation.

As has been mentioned before, under this regulation the future of resource management will be left to as yet undefined multi-species (MSTACs) and/or multi-annual (MATACs) TACs. How the Community envisages the application of MATACs and MSTACs remains a mystery. It could be that MATACs are to be set for a certain TAC of fish for a number of years; or maybe, while TACs will be set for a number of years, some adjustment of the TAC will be permitted, according to catches or to developments in fish populations or in the fishing industry. For example, the Commission's Scientific and Technical Committee for Fisheries (STCF) interpreted MATAC in its March 1992 meeting [182] as a system of borrowing quota from the following year to allow fishing to be continued. These quotas have to be paid back as a lower TAC allocation in later years.

Regarding the implementation of MSTACs it is likely that these will be set for species that are usually found in mixed fisheries or related to multispecies community interactions, for example predator-prey interactions and living in the same environmental habitat. Under such a scheme quotas for one species may be swapped for another with some conversion rate or count against future quotas [182]. Another possible application would be setting MSTACs to the level of the most vulnerable species in mixed fisheries catches. In such a way fishing would have to be stopped before extensive by-catches were taken.

The possible contribution of the MATACs and MSTACs to the management of the resource is not free from dispute. With respect to the present techniques of stock assessment (discussed in the second chapter) the scientific forecast is valid for up to three years and it has a degree of satisfactory reliability only for the so called "analytical TACs". Thus , given in the Community the small number of analytical TACs, whose validity is disputable (Pope, 1992), the benefits of such a system of MATACs for the conservation of the resource are questionable. Likewise, the application of MSTACs as a conservation measure raises similar biological controversies. Although multispecies interactions are considered by fisheries scientists (Garrod, 1988) and there is a great deal of effort to understand and include multispecies approaches in models of fish population dynamics (Cook and Armstrong, 1986; Getz *et al.*, 1987; Koslow, 1989), for the time being the practical reality for fisheries management is that there is little to gain from the use of such poorly established multispecies approaches. If the Community is determined to apply MATACs and MSTACs it must also be determined to face the costs of data collection to meet the requirement of such multi-parametric models.

Even if the biological and ecological approaches to multiannual and multispecies fish population models were robust and well established, there is a series of economic considerations as to the effect of the MATACs and MSTACs on market forces and fishermen's behaviour which in turn might create undesirable side effects for the resource. For example, given the uncertainties about the exact current fleet capacity and the exact state of the stocks, when a certain level of quotas is allocated to fishermen for a number of years, there would be a strong incentive to fish the quotas as fast as possible in order to benefit from the short run fishing possibilities that the MATACs allow. Furthermore, with MSTACs the already acute problem of discards may be aggravated given the chance to discard low priced species for higher priced ones in order for fishermen to maximize their revenues. Thus, it is essential to introduce strict control or even forbid discards under this system, as Norway does for a limited number of species. How discarding can be effectively controlled or banned is not to be learned from Norwegian practice, since the Norwegian authorities usually take action only when excessive quantities of small fish are landed. Discarding still remains a great problem for Norwegian fisheries [176]. Therefore, there is a number of factors that have to be taken into consideration before the application of MATACs and MSTACs. Such additional measures could possibly cause more socioeconomic upheaval in an already suffering industry with ambiguous results for the state of the resource. The industry does not only suffer from economic, social and resource mismanagement but also from a plethora of Community regulations and measures that take a lot of effort to interpret and comply with.

6 The Common Organization of the Market

The Common Organization of the Market in fishery products was established, as mentioned above, on 20 October 1970 by the regulation 2142/70 [11]. It contains the same broad principles which the Treaty of Rome introduced for the Common Agricultural Policy in articles 39-43. This regulation was amended after the 1972 Act of Accession (articles 39-43) and altered again in 1976 [57] and in 1981 [58]. Part of the 1981 regulation, except for articles 13 and 14 that contained some pricing provisions, has applied since 1 June 1982. It was designed to rectify that of 1976 and to cope with the changes in the law of the sea after the introduction of the 200 miles EEZ. The whole regulation was applied on 1 January 1983 [59]. This basic marketing regulation of 1981 was further amended by the Spanish and Portuguese Act of Accession [43]. At the end of 1991 [73] the Community produced an improved, comprehensive regulation on the common organization of fishery products which to a great extent re-stated the basic 1981 regulation. Finally, only after a year of the operation of this regulation the Community realised that, because of the number and complexity of amendments and the lack of clarity and rationality essential for such rules, the 1991 regulation was in need of replacement.

The new marketing regulation at the end of 1992 [72] is to a great extent a repetition of that of 1981 but includes some new provisions, for example for certain fishery products like sardines, tuna and bonito, pricing provisions for certain frozen products and a new storage aid scheme that aims to follow market developments and changes in fishing activities in the recent years. This new regulation, although it is too early to assess its effects, seems more integrated and comprehensive compared with the previous regulations. It requires the synergy and the co-operation of the Member States. Article 26 assures equal access to ports and first stage marketing and technical installations for all vessels flying a Member State's flag in any E.C. port.

Contrary to other products in other sectors of the economy, fishery products and certain agricultural products are controlled and managed in the international market by the application of customs tariffs. They are never left to the free and unrestricted operation of market forces. In 1962, at the end of the Dillon Round of GATT negotiations, the Common Customs Tariff was initially drawn up for fishery products. This restricted the Community's flexibility of manoeuvre in devising market management mechanisms for its internal and for its international market with third countries. The reason for this treatment is due to two main factors. 1)The fishery market is usually characterised by seasonal fluctuations in supply and 2) there is a large number of fishermen employed in small scale fishing unable to control prices and, therefore, their income.

According to the new marketing regulation of 1992 [72], the competence of the Common Organisation of the Market for fishery products covers the following:

1   fresh fish (live or dead), chilled or frozen;
2   dried fish, salted or in brine; smoked fish, whether or not cooked before or during the smoking process;
3   crustaceans and molluscs, whether in shell or not, fresh (live or dead), chilled, frozen, salted, in brine or dried; crustaceans in shell, simply boiled in water;
4   animal products not elsewhere specified or included; dead animals under the above, unfit for human consumption: fish, crustaceans and molluscs;
5   prepared or preserved fish, including caviar and caviar substitutes;
6   crustaceans and molluscs, prepared or preserved;
7   Pasta products (e.g. lasagne, canelloni, ravioli etc.), whether or not cooked or stuffed, containing more than 20% by weight of fish, crustaceans, molluscs or other aquatic invertebrate.
8   flours and meals made by fish, crustaceans or molluscs, unfit for human consumption.

The fish oils are not included in the above, since they are classified as oils and fats under the regulation 827/68 [60].

The components of the Common Organization of the Market in fisheries products

The provisions of the Common Organization of the Market in fisheries products include the following:

1 common marketing standards;

2   producers' organizations;
3   common prices;
4   a system governing trade with third countries.

Common Marketing Standards

With regard to the internal organization of the market the marketing standards are the only compulsory element. These are of two types according to article 2 of the new marketing regulation in 1992 [72].

The "classification" standards are designed to facilitate a uniform, common, prices policy throughout the Community. They determine the class to which the product belongs and are applied to fresh products at first-sale on the Community market. They have to meet certain quality specifications (size, freshness etc.). It is only a logical supposition, not yet written in the Community law, that these standards come after resource management and conservation policy. This stems from the argument that everything which is legally fished must also be able to be sold and, therefore, the marketing standards must comply with the resource management rules. Alternatively, the common marketing standards on the minimum size of fish for human consumption reinforce the conservation policy of the Community.

The "composition" standards apply to processed products. These products must comply with presentation and packaging standards which are designed to ensure fairness in transaction and market transparency. In recent years standards for the canning industry have been laid down, particularly for canned sardines in 1989 [61] and for canned tuna and bonito in 1992 [62].

Under article 3 of the 1992 regulation [72] the common marketing standards are liable to inspection by the Member States and infringements are to be penalised. This provision was included also in the basic 1981 regulation [58] although it was not adopted until December 1985 [67]. Since then the Community's standards have been criticised for their complexity and for the difficulty in checking and observing them [68]. For example, cod has to be graded into three categories of freshness and five of size, giving 15 possible grades that not only add to the complexity of the system but also to the cost of production with ambiguous benefits for the consumer.

There were no physical and health standards until 1991. A Council directive of 22 July 1991 [63] lays down physical and health conditions for transportation on board and other means of transportation, processing, preservation packing and placing in the market of seafood. This directive includes provisions for the internal market as well as for imports from third countries and in both cases provisions have been made for the establishment of veterinary checks on the fish products. All these measures aim to establish uniformity in the Member States' health law for these products, as border controls were dismantled on 31 December 1992. However, the market in fish did not comply completely with the provisions of the directive, because of shortage of time or because of waves of demonstrations in Member States where the measures were considered extreme. Therefore, the Council adopted transitional measures to facilitate the change over to the arrangements provided by the directive. In the case of Member States the transitional period runs from 1.1.93 until 31.7.93 [64] and in the case of third countries' imports the transitional period is set from 1.7.93 until 31.12.94 [65]. Despite the transitional measures, the Community is determined to set physical and health conditions for fishery products and to certify the consistency and uniformity of seafood sold on the market. This is obvious also from the recent Commission's decisions in 1993 [66] concerning the quality of raw and cooked crustaceans, molluscs and shellfish that are placed on the market.

Critique of the Common Marketing Standards

The justification for the establishment of the Common Marketing Standards can only rest on the existence of the complicated structure of the Common Marketing policy. This is because the existing operation of the common price mechanism requires a confusing nomenclature [139] of a variety of classified and standardised fishery products. The whole system of intervention for fishery products (see below) is applied to fish which conforms to Community marketing standards.

It is hard to find any other justification for the existence of these common marketing standards, though a case can be made for the need for quality and health standards for the benefit of the consumer. Nevertheless, the consumer of average education is nowadays aware of healthy eating, of a balanced diet and of the impact of various chemical additives in food. Therefore, the establishment of quality and health standards for fishery products by the Community is unlikely to make any difference to the demand for these products. It is ingenuous to believe that the standards that are set by the Community, for example, for the canning industry are those that force this type of industry to comply with the taste of the consumers. There is little sense in establishing marketing standards for a product if the consumer is not ready to accept and buy this product. A better way to persuade the consumer to buy a product that meet certain standards is by promoting the benefits he or she could have. Therefore, the objectives of laying down common marketing standards for fishery products such as "to improve the quality of fish marketed and to facilitate their sale" are somewhat confusing and contrary to forces of the market economy. There is little benefit for the consumer or the producer from the establishment of a variety of complicated categories of freshness and size for fish. The net result of such measures is that the consumer pays more for the good and the producer, restricted by law, fails to meet the real demand of the consumer.

Another objective of common marketing standards is to produce a common price policy throughout the Community, which by nature is not uniform. There are several reasons for this. They may lie in the differences in quantities of supply, consumer tastes throughout the Community, the variety of species distribution and in the Community's trade policy. Which of these is the main reason is questionable, but it is not clear at all that the legal enforcement of standards for fishery products can create uniformity of prices all over the Community or that this uniformity would be desirable. It is a fact that the value per tonne of all fish products in ten Member States - for Netherlands and Portugal there are not reliable data - in 1990 varies considerably from one Member State to another (Figure 6.1). This is quite natural because the Member States do not supply to their national market the same quantities of fish products and the consumers tastes are not uniform. Therefore, by trying to create a uniform demand with the enforcement of common standards for fishery products, this leads not only to demand alterations but also to changes in supply, which both cause a great distortion of the market in fish.

Producers' Organizations

The new marketing regulation of 1992 [72], particularly articles 4-7, creates a legal exception to the principle which prohibits cartels under the Treaty of Rome by establishing producers' organizations for fishery products. Producers' organizations are subject to limits; article 4(1) reads "any recognised organization or association of such organizations is established on producers' own initiative for the purpose of taking such measures as will ensure that fishing is carried out along rational lines and that conditions for the sale of their products are improved". In order to obtain recognition from the Community a producer organization must fulfil certain conditions that are contained also in the basic marketing regulation of 1981 [58] supplemented by a regulation in 1980 [69] and subsequently amended by others in 1984 [70] and in 1987 [71]. These in turn are re-emphasised in the recent 1992 [72] regulation. The conditions for a producer organization include:

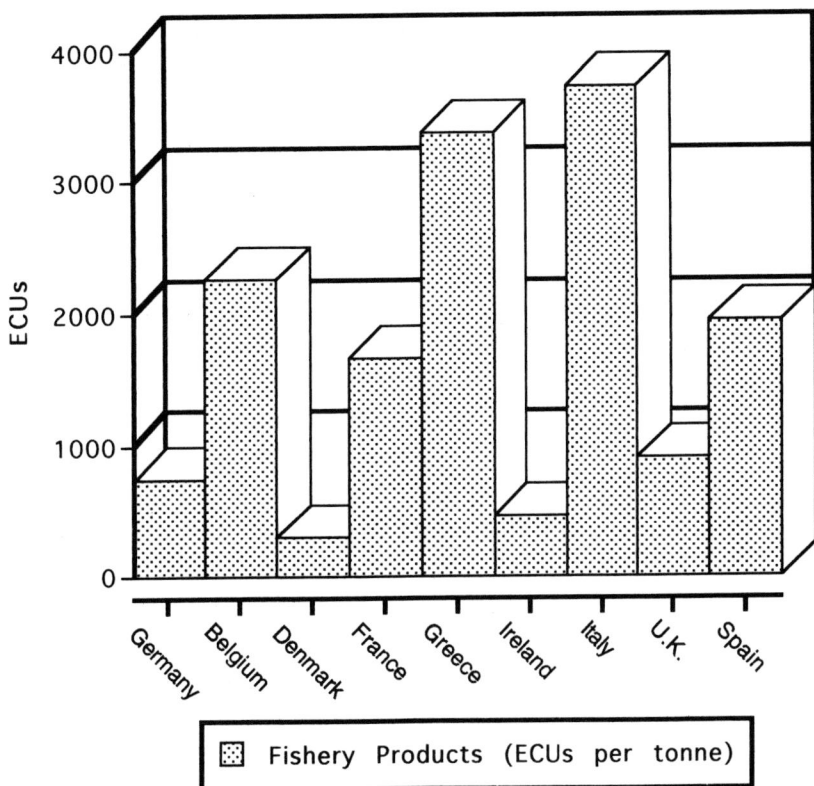

Figure 6.1 The average landing price per tonne for fishery products in ten E.C. States in 1990

source: D.G. fisheries XIV-A-3.(30.6.1993)

1 It must be recognised also by its national legislation.
2 Its members must be professional fishermen and its activity must cover fresh, chilled or frozen fish dried, salted, smoked or fish in brine, or crustaceans, or molluscs.
3 Its members must be able to join and leave freely, if they wish to do so. In the case of a member wanting to leave, he or she must have been a member for three years and give a year's notice.
4 Its members must dispose of their catches through the organization. They must apply rules aiming to improve quality, to adapt the volume of supply to market requirements and to ensure the proper management of the catch authorised by the quotas.
5 They must not discriminate between Community fishermen within "an economic area" on the grounds of nationality or place of establishment.
6 Its members must not hold a dominant position on the common market by obtaining a monopoly and applying monopolistic rules unless the objectives in article 39 of the EEC Treaty, mentioned above, are pursued.

The formation of producers' organizations is not compulsory. However, fishermen, who do not belong to such organizations and market fish products, are under the control of Member State regulations that aim to ameliorate disrupting influences on price formation and to comply with producers' organizations in improving quality and ensuring proper management of the quotas authorised. These rules, laid down by Member States, are left to their legislative will, i.e. whether or not to extend regulations for fishermen, who are non-members, to certain categories of catch products and sales. Such arrangements must not lead to an infringement of E.C. marketing rules and Member States must take all the appropriate measures to penalise breaches of these rules. In addition, according to the article 5(4) of the new marketing regulation of 1992 [72] Member States may impose fees on non-members of producers' organizations to contribute towards the administrative costs resulting from the application of the price system, which will be analysed below. They concern market withdrawal and carryover premiums for fish products. The Member States that have introduced such a system for non-members of producers' organizations may indemnify them for any fish not marketed, or withdrawn from the market up to 60% of the withdrawal price.

Although the formation of producers' organizations is not compulsory, it is encouraged by the Community by means of financial assistance. Under the article 6 of the 1981 regulation [58] Member States may make aid available to producers' organizations "to encourage their formation and to facilitate their operation". For the application of this provision detailed rules are contained in several other subsequent regulations [74]. The same provisions are also contained in the 1992 regulation [72] under the article 7 for the producers' organizations that are established after 1 January 1993. According to these provisions aid may be granted on a decreasing scale and within certain limits [75] for three years following the recognition of the organization within five years of the date of recognition. Under article 25 of the 1992 new marketing regulation [72] Member States are to be re-imbursed for half of this aid from the Guidance Section of the European Agricultural Guidance and Guarantee Fund. In addition to this aid, article 7(4) of the 1992 regulation provides that for five years following the creation of the intervention funds to finance producers' organizations' withdrawal arrangements, a Member State may make aid available either directly or through credit institutions in the form of loans on special terms to cover part of the anticipated cost of market intervention. This aid must be notified to the Commission each financial year. Between 1983 and 1990 the aid given to producers' organizations by the Community was 420,000 ECUs, which represent a very small percentage, only 0.02% of the Community expenditure for fisheries and 2.2% of the Community's expenditure for the Common Organization of the Market in fisheries products [76]. In the following years the Community made a decisive move to encourage the formation of producer groups in the fisheries sector. This change is obvious from the aid given to producers' organizations. In the years 1991 to 1993 the aid to producers' organization increased from 477,000 ECUs in 1991, to 600,000 ECUs in 1992 and 1 million ECUs in the 1993 budget [77].

The Commission publishes from time to time a list of producers' organizations recognised by the Community. The most recent list was published on 21 February 1990 [95] with a total number of producers' organizations in the E.C. of 144. There are 1 in Belgium, 3 in Denmark, Ireland, the Netherlands and Greece, 12 in Portugal, 14 in the UK, 19 in Germany and Italy, 30 in France and 37 in Spain. In all the Member States, except in Belgium and the Netherlands, there are separate organizations for the different sectors of the industry. For example, there are separate organizations for offshore, high-sea, local inshore, deep-sea and other fishing. The fish production covered by the above producers' organizations represents about 60% of the total fish production in the Community [54].

The price system of the Common Organization of the Market

The price system of the Common Organization of the Market is concerned with the formation of price at the first sale of fish, from the producer (fishermen) to the wholesaler or retailer and it is not concerned with the formation of price in any of the subsequent stages of marketing. Most first-hand sales of fish for human consumption take place through auctions in the E.C.. The price system for fishery products is as follows.

Guide price Before the beginning of each fishing year the Council lays down a guide price for each of the main species valid for the whole Community. The guide price is based on the average of the actual market prices for these products during the three fishing years immediately preceding the year for which the price is fixed and on an assessment of supply and demand prospects, producers' incomes and consumers' interests.

Withdrawal price For the major species a withdrawal price is fixed below which producers' organizations cannot sell products supplied by their members. The withdrawal price is either fixed by the Community on the basis of the freshness, size or weight and the presentation of the product, at an amount equal at least to 70% and not exceeding 90% of the guide price. Withdrawal prices can be also set by producers' organizations, when they feel that the market price needs additional stabilisation. These prices may fluctuate, within a margin of tolerance extending 10% below and 10% above the Community withdrawal price. Other withdrawal prices set by producers' organizations for certain fishery products (i.e. fresh or chilled products of sole, pollack, blue whiting, horse mackerel etc. are found in the Annex VI of the 1992 regulation [72]) for each year and are called "autonomous withdrawal prices". These must not exceed 80% of the average price of the products in question during the previous three years and they cannot fluctuate more than a 10% band above or below during the year. The autonomous withdrawal prices are set by the producers' organizations in an attempt to remove the seasonal fluctuations in market prices for the specified fishery products. They cover fish products demand for which is lower than for other fishery products covered by the Community withdrawal price and which are subject to regional price disparities.

Financial compensation is granted to producers' organizations only where products withdrawn from the market are disposed of for purposes other than human consumption or in such a way that they do not interfere with the normal marketing of these products. Detailed rules for granting financial compensation to producers' organizations can be found in the recent 3902/92 regulation [83]. To finance withdrawal measures, producers' organizations create the so called "intervention funds" fed by contributions that for certain species are financed by the Community from the Fisheries Guarantee Fund, according to strict rules. The level of compensation is inversely proportional to the volume of catches withdrawn from the market. According to the 1992 regulation [72] the range of compensation is from 87.5% of the withdrawal price for quantities withdrawn that do not exceed 7% of the annual quantities which are put up for sale, 75% for quantities which exceed 7% but not 14% and the Community's contribution is equal to 0% for quantities that exceed 14%. In the case of those products, to which autonomous withdrawal prices apply, the aid given by the Community is at a flat rate of 75% of the autonomous withdrawal price for quantities not exceeding 10% of the annual quantities put on the market.

This system of co-responsibility for fishermen in the financing of the withdrawal measures is designed by the Community to make producers aware of their responsibilities and encourage them to curtail their catches according to market demand. This aid is not payable to individuals but only to producers' organizations which are responsible for its distribution.

Carry-over aid The aim of this system is to give incentives to the producers' organizations to carry-over fish of a high commercial value, which have been withdrawn from the market, and put them back on the market again latter, instead of destroying them or reducing them to fish meal and oil. The fish products that qualify for this carry-over aid, according to article 14 of the 1992 regulation [72], are those in Annex I A, D and E that are supplied by a member producer, are withdrawn from the market at the withdrawal price or in the case of fish products listed in Annex I(E), where a buyer has not been found at their Community selling price [78]. A margin of tolerance has been set equal to 10% above or below these prices to take account of seasonal fluctuations in market prices. The aid granted for these products may not exceed the amount of the technical and financial costs associated with operations which are essential for processing and/or storaging quantities not exceeding 6% of the annual quantity of these products put up for sale. A recent regulation of 23 December 1992 [82] sets specifically the processing methods that these products must undergo and gives details for their storage conditions. It also demands that the maximum period of storage of these products before they are put back on the market must not exceed six months.

Private storage aid In an effort to limit the value of products that are withdrawn and then destroyed, storage premiums are granted for certain products frozen on board [79]. Private storage aid is to be granted when the market price remains for a "significant period" below 85% of the guide price. The aid may not be granted for more than 15% of the quantity offered for sale and the aid may not exceed the sum of technical costs and interest for a maximum period of

155

three months. These products must be stored for three months' maximum period and then placed on the Community market.

Aid for tuna for the canning industry   Under articles 17 and 18 of the 1992 marketing regulation [72], as well as in its predecessors [58,73] to a lesser extent, provisions were made for aid to producers of tuna (of all species) intended for the canning industry. The Council each year fixes a Community producer price which is calculated in much the same way as the guide price for other fish products. Compensation is granted to producers' organizations for quantities of tuna species caught by their members and sold and delivered to canning industries located in the Community. It is granted if the quarterly average price in the Community market and the free-at-frontier price (see below) plus any applicable countervailing charge are lower than 93% of the Community producer price. Such compensation must not exceed, i) the difference between a "triggering threshold" of 93% of the Community producer price and the average selling price of the product, ii) a flat-rate amount equivalent of 12% of the above "threshold" price or iii) the difference between that "threshold" and the average selling price in the Community market for certain quantities of fish.

In conclusion, the provisions focusing on tuna for the canning industry are the newest component of the 1992 regulation [72] in relation to its predecessors [58,73]. For these provisions, the whole regulation has been welcomed by the Economic and Social Committee [80] in its opinion on the regulation. The Community in laying down this aid scheme aims to ensure the income of the producers of this product which is threatened by a fall in import prices for tuna for the canning industry.

The collapse in the world price of tuna arose from a world-wide reluctance to buy tuna caught by methods, which simultaneously endangered dolphins [108]. The loss of normal markets and the fall in price caused the tuna to be off-loaded into the E.C. market with adverse effects on indigenous fishermen. However, the USA and not the Community had the initiative to bring pressure on the world's largest producers (mainly Mexico) through an embargo of all tuna imports unless they are fished in a "dolphin safe" way

The Community in an effort to offset the disadvantages that the producers face from cheap imports is bound to submit a report on the situation of the tuna market with appropriate proposals before 30 June 1994.

Critique of the intervention scheme for fishery products

The establishment of producers' organisations by the Community represents another violation of the market economy for fisheries. The setting up of such organizations could be one of the most effective ways of protecting the producers' interests if they were established on a different basis. The objectives of these organizations, as the Community envisages them, are oblique and comprise the easy way out, avoiding firm and defined objectives. The regulation that establishes them is full of notions and words that can be translated as one pleases in order to serve better one's interests. For example, there is not a clear definition of "fishing along rational lines" (article 4(1) of the 1992 marketing regulation) while this has to be ensured according to the regulation, by such organizations. In addition, according to the regulation, among the aims of producer organizations is that to "improve their product quality" as if producers needed such a ruling to comply with demand, and "to adapt the volume of supply to market requirements", as if fishermen never before had such an aim.

Another point about the Community's regulation for producers' organizations is that they are to be set up on a so-called voluntary basis. This, up to a point, is true. Individual fishermen can freely join a producer organization; however, if they wish to leave they have to engage in a bureaucratic procedure that may be designed to abate any possible united activity. For example, in the case of a crisis in the fishing industry their members cannot show their dissatisfaction with the system by simply leaving the organization. In order to leave they have first to be members for three years and give a year's notice.

Although some producers' organizations in some Member States (e.g. U.K. and Netherlands) are involved in buying and selling quota entitlements, fish selling, processing, chandlery and oil supply, such activities are not required by any E.C. regulation but result rather from their own initiative. Therefore, the role of producers' organizations as the Community conceives them at the moment remains largely administrative, although it has been envisaged that they should have a wider role in controlling fishing effort and in decommissioning (Slaymaker, 1992). Perhaps in the future under a new CFP all producers' organizations will be able to purchase their quota entitlements and licences and play an active role in reducing overcapacity.

The decision of fishermen to form these organizations presumably relies on balancing the benefits against the disadvantages that may arise from the formation of such a union. These disadvantages comprise the payment of fees for membership which are intended to cover the administrative costs resulting from the application of this system. The administrative work includes the calculation of compensation according to the provisions of the intervention scheme. In doing this, producers' organizations' officials have to follow a confusing nomenclature of fishery products, comply with the marketing standards and keep detailed records of the methods used for these calculations.

At the moment the major incentive for a fisherman to become a member depends on the quantity of financial assistance given by the Community. This financial assistance takes the form of aid to encourage the formation of producers' organizations and others that are derived from the operation of the intervention arrangements for fishermen members of producers' organizations. Since 1983, when the regulations giving such financial aid became fully operational, the amount of financial assistance has grown (Figure 6.2).

Unfortunately, the effect of the intervention scheme on fish price formation cannot be established. In figure 6.3 the average landing price per tonne of all fisheries products in the E.C. is shown during the years 1973 to 1990. There is a sharp increase in price (Figure 6.3) and fish landings (Figure 6.4) in 1986 which can be partly attributed to the two new Member States, Spain and Portugal whose catch of high valued species (2,140,000 tonnes in 1986) represents almost half of the catch of all the other Member States and almost the same proportion in terms of first-hand sale value at that period of time. Since then, the amount of fish landed has decreased but it was also followed by an increase in imports while exports remained stable (Figure 6.4). On the other hand fish consumption over the years for all the twelve member States since 1970 does not appear to follow a particular trend (Figure 6.5). In addition data for fish consumption must be treated sceptically particularly since 1986 as there is some inconsistency among different sources (e.g. FAO Yearbooks of fisheries statistics, Eurostat basic statistics and Agriculture statistical yearbooks). It must be noted here that the fall in consumption of fish in 1989 (figure 6.5) does not conform with the supply in figure 6.4. Therefore, it is difficult to establish the demand inside the E.C.. There are many reasons which can account for the fluctuations in demand and supply of fish; for example, increased fishing costs and stock depletion can affect the supply, and the rate of inflation and the overall growth of the economy in Member States can affect the demand. The sharp increase in the Community's expenditure under the intervention scheme in 1988, that was almost three times the amount spent in 1987 (Figure 6.2), is less than 0.001% of the landing value of fish in that year. Overall, If one considers the low percentage (less than 0.001%) of the Community's expenditure under the intervention scheme in relation to the value of landings during the years of the full implementation of this scheme between 1983 to 1991, it is unlikely that this can contribute to the price formation of fish. It is not clear from the available data if the intervention scheme contributes

ECU million

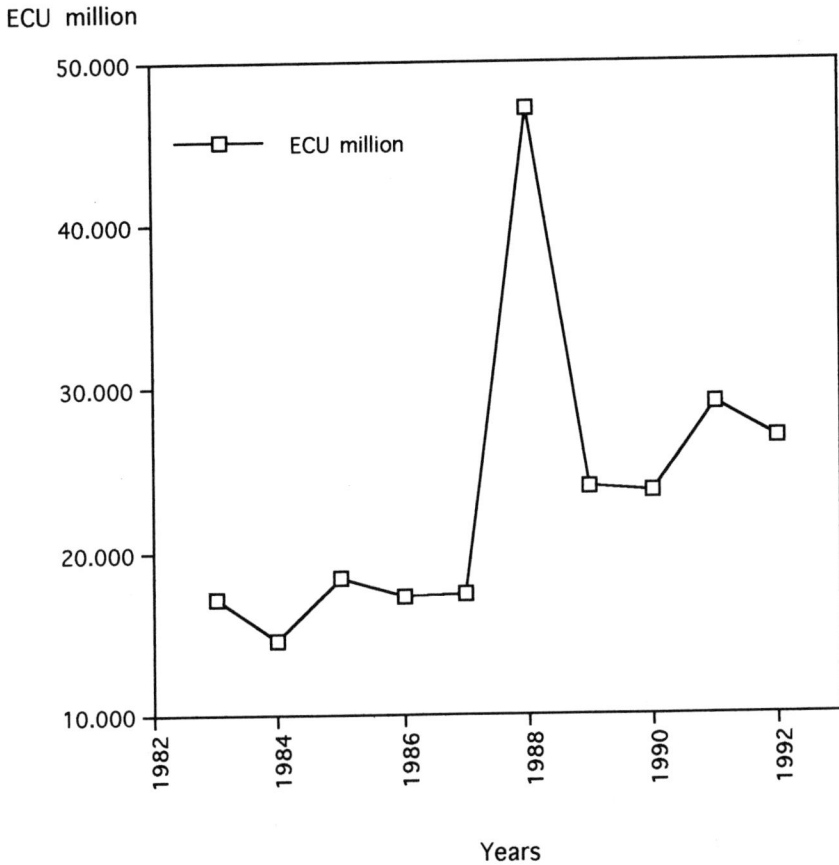

note: Data1983-1985 for 10 Member States, data 1986-1992 for 12 Member States.

Figure 6.2 Community expenditure under the intervention scheme of the Common Market Organization for fishery products

source: SEC(91) 2288 final (18.12.1991) and O.J. (8.2.1993) L31

159

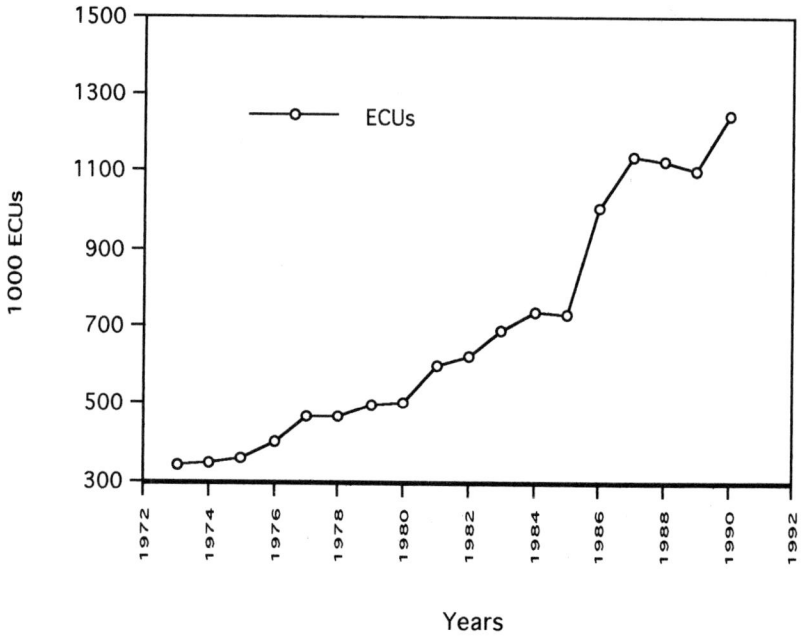

note: Data 1973-1980 for 9 Member States, data1981-1985 for 10
Member States, data 1986-1992 for 12 Member States

Figure 6.3 Average landing price per tonne of all
fisheries products in E.C.

source: D.G. XIV, XIV-A-3 (30.6.1993)

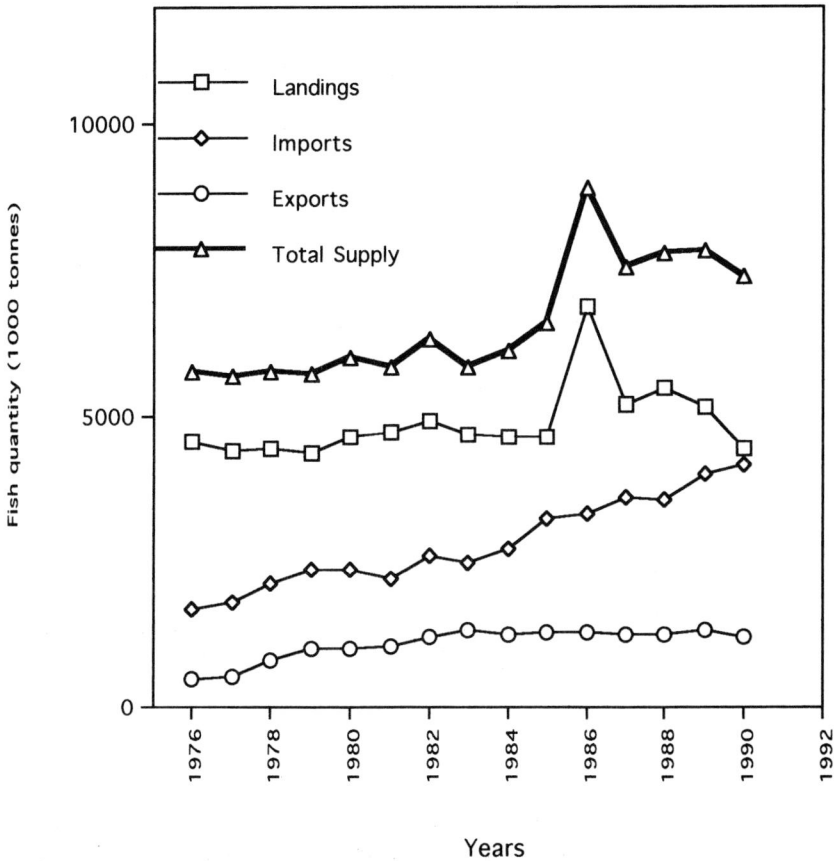

note: Data 1976-1980 for 9 Member States, data1981-1985 for 10
Member States, data 1986-1992 for 12 Member States

Figure   6.4   Supply   of   fish   in   the   Community

sources: D.G. XIV, XIV-A-3, 30.6.1993, SEC(91) 2288 final, 18.12.1991
and Review of Fisheries in OECD Member Countries, Paris, 1993

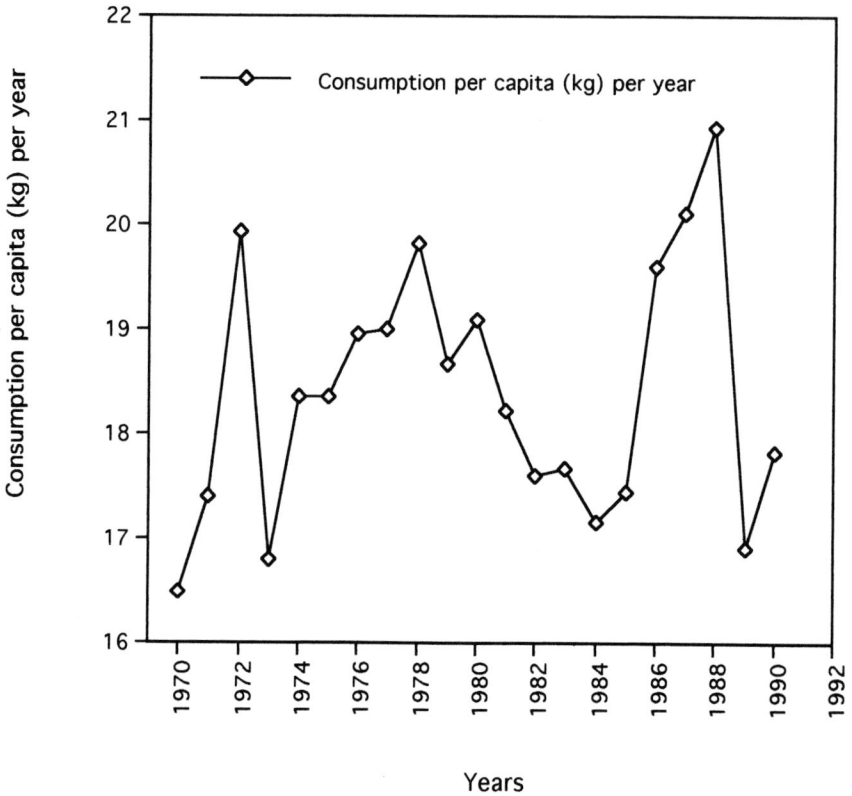

note: Data 1970-1972 for six Member States, data 1973-1980 for 9
Member States, data 1981-1985 for 10 Member States, data 1986-1992
for 12 Member States

Data for fish consumption must be treated sceptically particularly since
1986 as there is some inconsistency among different sources, e.g. FAO
Yearbooks fisheries statistics, Eurostat (1A) basic statistics and Agriculture
statistical yearbooks

Figure 6.5 Consumption flow of fish products per capita in the E.C.

source: D.G. fisheries, XIV-A-3

significantly to price formation for fish in the Community and if it has attained its general objective i.e. (article 39 of the Rome Treaty) of ensuring availability of supplies for consumers at reasonable prices.

One of the main conditions for the formation of producers' organizations under the new marketing regulation in 1992 [72], as well as in its predecessor [58], is that fishermen members must not hold a dominant position on the market by obtaining a monopoly unless the objectives of the article 39 of the EEC Treaty (page 91) are pursued. These objectives have been set to ensure a socially desirable operation of the market in fishery products. In welfare economics there is a basic doctrine that a monopolist performs in a socially undesirable manner. Using conventional fisheries economics, we assume that the supply curve is backward bending and the marginal cost rises asymptotically towards the line Quantity $= Q_{MSY}$ (see first chapter) and for simplicity that the demand curve is linear (Figure 6.6). Social welfare is maximised at the point where the demand equals the marginal cost. This point is a socially efficient point of production since it maximizes the net difference between the social utility of consumption and the total cost of production. In a monopoly, a monopolist's profit is maximized when marginal cost equates marginal revenue. It would be possible to assume a situation where: 1) the long term backward bending supply curve in an open access market for fisheries exists, 2) that fisheries firms and fishermen owners of fishing vessels in an open access market are taken over by a cartel 3) that the short run output would be unchanged since the same means of production would be employed and 4) the state of the fish population remains the same during this time. Then, the output of the variety of firms and vessel owners in open access would become the output of a monopolistic industry and, therefore, we can assume that in the short term after the monopoly is established the supply conditions would remain the same.

It is now possible to compare the monopolistic (producers' organization) equilibrium position with the socially optimal output. The output is smaller ($0Q_M < 0Q_o$) and the price higher ($0P_M > 0P_o$). The net social benefit declines from EHF to EHGM The policy of encouraging the formation of producers' organizations is inconsistent with that of maximizing social welfare from fisheries. It is unlikely, however, that in devising fisheries policy any attention was paid to the conclusions of fisheries economics. In practice, it is likely that cost and demand conditions will change when a monopoly takes over a competitive industry (Baumol, 1977) and furthermore, it is often argued that the discount rates in natural resource situations in the private sector tend to be greater than socially desirable rates (Solow, 1974). The shape of the supply and the marginal cost curves would need to be amended for these reasons.

163

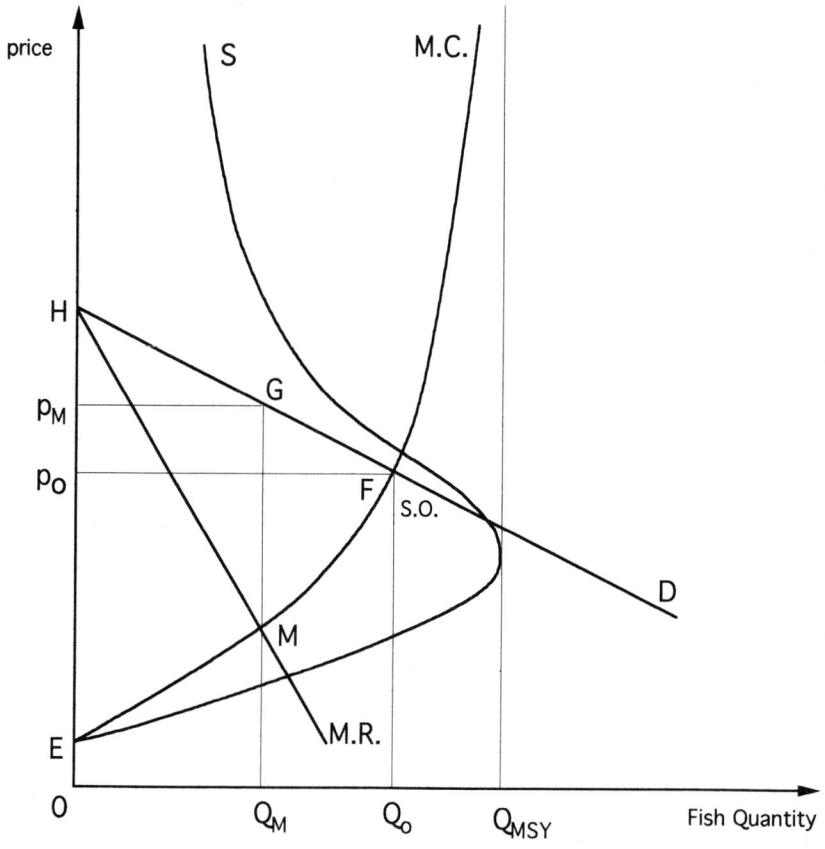

Figure 6.6   A theoretical comparison between
monopolistic and socially optimal
output in fisheries

However, regarding the competitive market in Member States, as well as the above condition for the formation of the producers' organizations and the objectives of the Community in article 39 of the Rome Treaty it is not clear how the Community envisages the situation. This is because such Community legislation is not backed by clear definitions of what "dominant position in the market" means in practice and there are no defined, or even envisaged, limits where the objectives of the article 39 which have been set to ensure a socially desirable operation of the market in fishery products could apply and work efficiently in practice. The above analysis although based on simplified assumptions reveals the contradictory elements in the above mentioned E.C. regulations concerning the objectives of establishing producers organizations in the fishing industry.

Furthermore, under the assumption of a linear demand curve for fish, where the middle point of this curve denotes unit price elasticity of demand it can be argued that although it may be rational for an individual fisherman to maximize his catch since one vessel's landings is likely to be a small part of all landings and, therefore, have little effect on price, for a producer organization which can attain some form of a monopoly power this might not be the case. That is because in a monopoly, producing a greater quantity beyond the middle point of the demand curve wll result in diminishing total revenue because of price inelasticity of the demand to the right this point. However, in practice such monopolistic power for a producer's organization is unlikely to be common, because there are several producers organizations landing their catches at the same port and also there is the disruptive influence of non-members. Although such an analysis is based on the simple assumption of a linear demand curve it elucidates the lacunae and the contradiction in E.C. regulations. These deliberately establish producers' organizations with envisaged monopolistic powers but at the same time the same regulation prohibits them from attaining a dominant position unless this is consistent with the socially desirable operation of the market in fishery products.

If the fishermen, who are not members of producers' organizations go their own way they would not be allowed to do so. Carefully designed regulations by Member States control any disrupting influences on price formation and ensure the compliance of non-members to Community legislation that covers the members of producers' organizations. There is, however, a united representation of all producers' organizations in the Community through EUROPECHE which makes its own recommendations for amendments to the CFP. However these recommendations implicitly accept the framework of the CFP and seek merely to improve the financial assistance from the Community and to relax the administrative burden.

The effect of the producer's organizations, the intervention arrangements and indeed the whole application of the marketing regulations on the volatility of landing prices of fish over the years will now be explored. There are no consistent data given by the Community on the monthly variation of landing fish prices in different Community ports or regions for a sufficient period of time before and after the full application of the intervention scheme. Therefore, consistent data on the monthly variation of landing price per tonne of selected species of fish from fisheries data on landings by U.K. fishing vessels in Scottish ports will be used. This will give some indication of the effect of the common marketing measures for fish on the degree of oscillation in the monthly landing price of fish in the Community. The six species of fish selected (Figure 6.7) are representative of those species, to which the provisions of the common marketing pricing arrangements for fisheries products apply according to the 1981 and 1992 marketing regulations. These provisions aim to act as a price stabilisation mechanism to ensure a "fair standard of living" for fishermen. Despite the full application of the intervention scheme for fisheries products in 1982, it can be established that for cod, haddock and whiting the monthly oscillations in prices were greater after 1982 than they had been before 1982 (Figure 6.7). For mackerel, herring and saithe, their monthly landing price volatility shows no effect from the application of the marketing regulation in 1982 in a comparison to the volatility of prices from 1975 to 1982. Therefore, although it can be argued that the fluctuations of monthly prices might have been even greater without the common marketing measures, the degree of volatility of landing prices of fish from Scottish fisheries data indicates that the common marketing system has not achieved the reduction of the price oscillations to the pre 1983 level for cod, haddock and whiting. Nor did it have any effect in restricting the price volatility for saithe, mackerel and herring to a point never before experienced in the pre 1983 years for these species.

However, compared to its predecessors, the 1992 marketing regulation [72] attempts to add more flexibility to the price mechanism by retaining the autonomous regional price and by establishing the carry-over and private storage aid to producers. In addition, provision is made for fishermen, who do not belong to or have departed from producers' organizations, and consequently from the Community intervention system. They remain entitled to the granting of an indemnity from Member States.

It is too soon to assess the pricing arrangements of this new regulation. However, for the arrangements that were included in previous marketing regulations, it can be assessed if they were effective in reducing the quantities of fish withdrawn. Nevertheless, the decrease in withdrawals is not their main

166

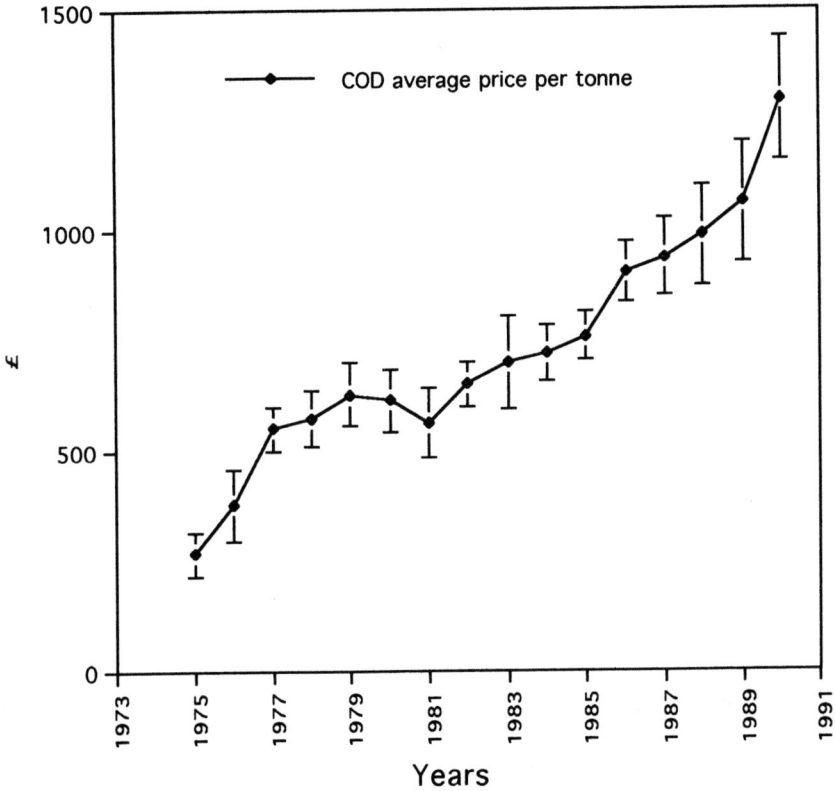

Cod

COD average price per tonne

Figure 6.7   Seasonal volatility of fish landing
price per tonne for six species landed
by U.K. fishing vessels in Scottish
ports.
Vertical bars represent Standard
Deviation

Haddock

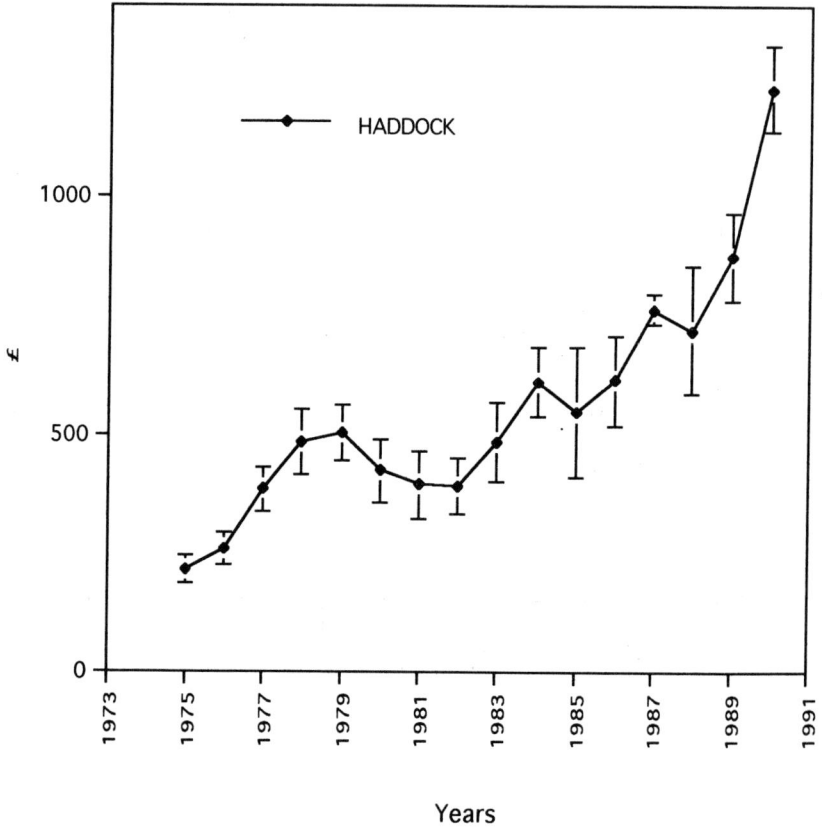

Whiting

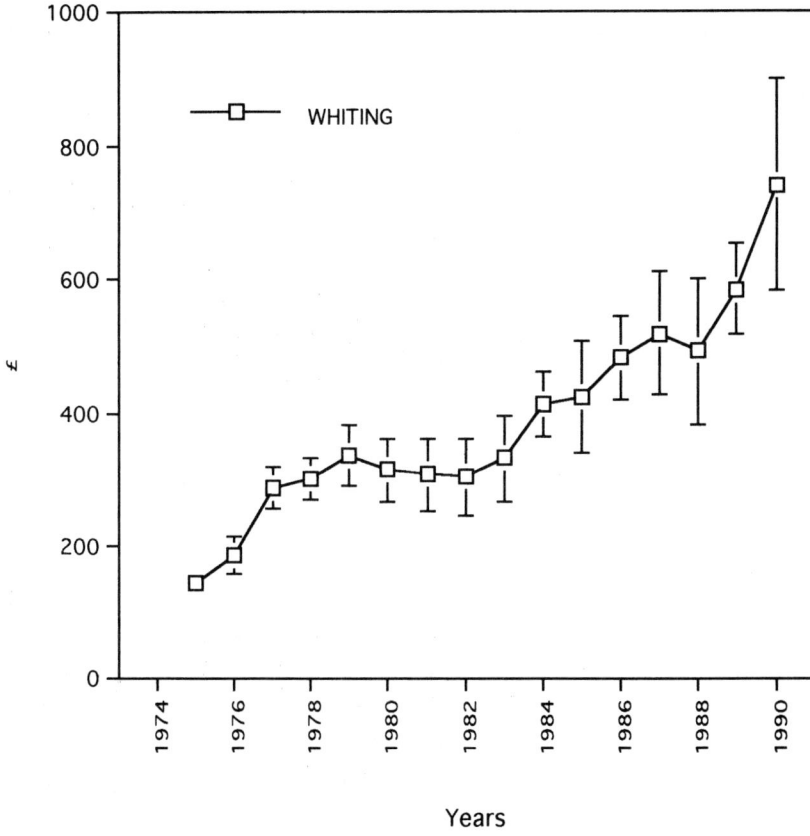

Saithe

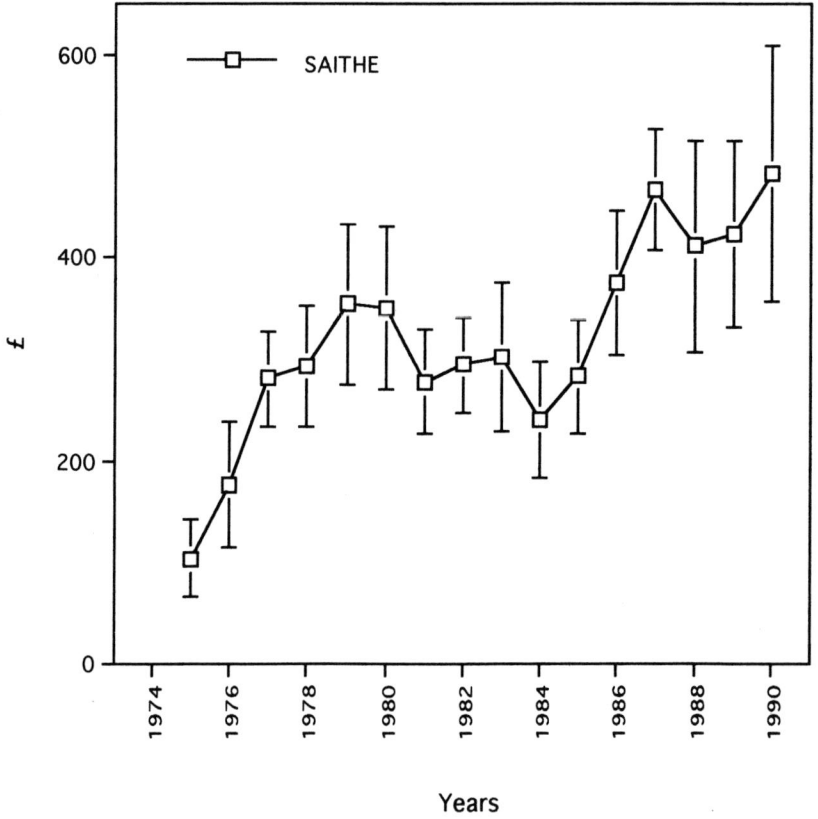

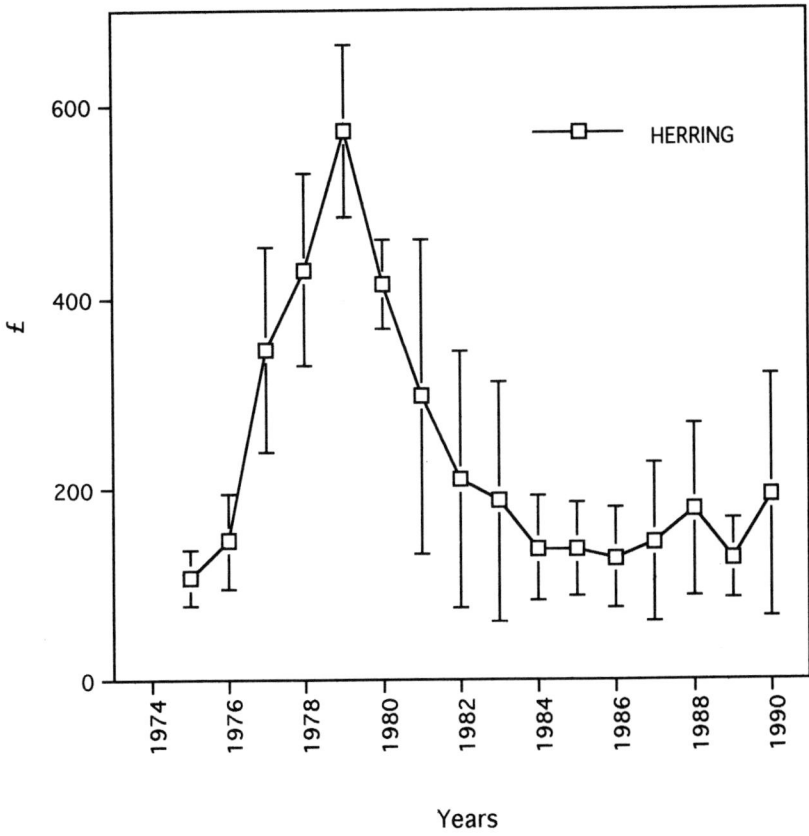

Herring

Mackerel

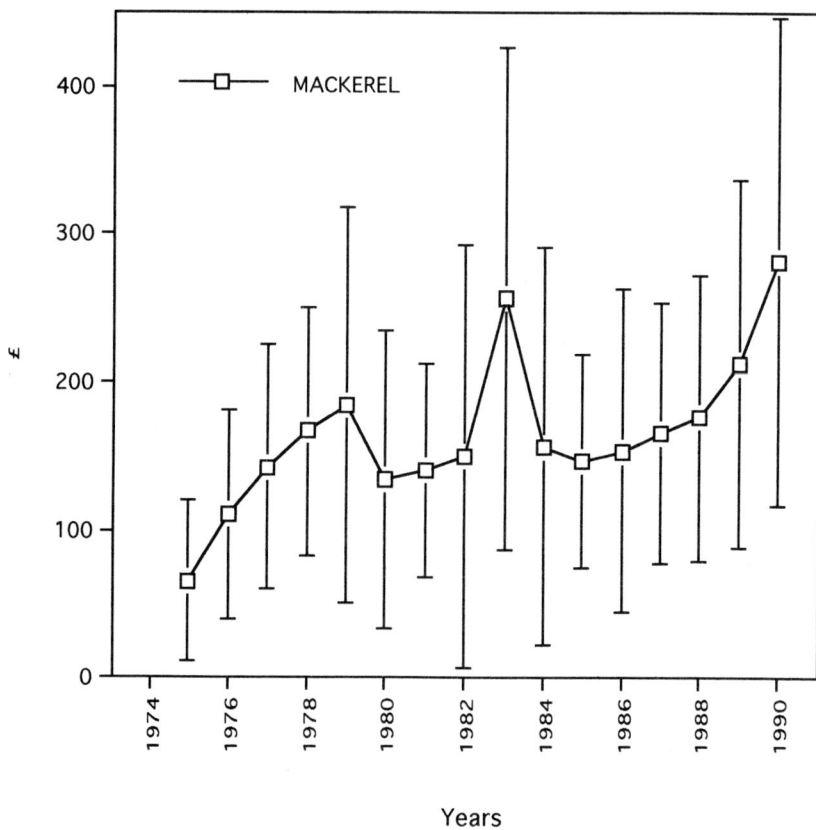

source: Scottish Sea Fisheries Statistical Tables, data from 1975-1990

172

objective. The aim of the intervention scheme is to protect fishermen's income. This is considered below. However, the decrease in withdrawals as a side effect can be assessed from the point of view that production resources are not wasted and the fish found a place on the market. According to the data in a report by the Court of Auditors [68], the effectiveness of the price mechanism, where the level of compensation decreases as the quantities withdrawn from the market increase, is considerable.

In the year 1982 (when the 1981 basic marketing regulation was supposed to be applied [58]) 139,000 tonnes of fish landed did not find a place on the market and were withdrawn. Whereas when the regulation applied, in 1983 and 1984 the amount dropped to 87,000 and 76,000 tonnes respectively. In 1988 the fish withdrawn were about 50,000 tonnes [84]. In 1983 and 1984 the reduction was not due to carry-over premiums because at that time the application of this system was in an embryonic stage. The decrease of the withdrawn quantities cannot be attributed to the intervention scheme with absolute certainty. This is because the intervention scheme does not cover all fish species and the uptake of the available funds is discretionary by the producers' organizations. In 1982, 4,900,000 tonnes of fish were landed in Community ports. For 1983 and 1984 fish landings in Community ports remained relatively stable (4,677,000 tonnes in 1983 and 4,638,000 tonnes in 1984). Demand remained relatively unchanged in 1982, 1983 and 1984, at 17.6Kg., 17.67 Kg. and 17.14 kg. of fish consumption per capita per year, respectively. However the price per tonne of fish rose considerably from 622 ECU in 1982 to 688 ECU in 1983 and to 734 ECU per tonne of fish in 1984. The higher withdrawn quantities in 1982 compared to those in 1983 and 1984 were due to the increased supply followed by a lower price of fish in that year. Thus it is not certain if the reduction of withdrawn quantities of fish in 1983 and 1984 compared to these in 1982 can be attributed to the intervention scheme of the Community alone, since fishermen could benefit more by selling fish as the price of fish was increasing during 1982 to 1984. Given the limited application of the price system to certain species of fish and comparing this situation to that in 1988 when landings increased (5,456,000 tonnes) and the demand for fish rose (20.94 Kg. per capita) as well as price per tonne of fish (1,122 ECU), it is more difficult to attribute the decrease in withdrawn fish to the intervention scheme. This emphasizes the fact that although the price system was designed to ensure support for fishermen's income its limited application to certain species of fish makes the system incapable of resisting market forces.

If all the parameters of production were stable the level of the withdrawn quantities could demonstrate if the guide and withdrawal prices are set at a socially acceptable level. This is because if they were set at a higher level more fish would be withdrawn from the market without necessarily an increased income for fishermen. As Young (1984) states, fishermen's income in respect to the level of withdrawal prices depends on the elasticity of demand. With a given demand and an increasing supply of fish the quantities withdrawn increase, with the effect that the compensation declines as the withdrawal price is related to the market price. In order to avoid this situation and limit the destruction of withdrawn catches, the 1981, and also the 1992, marketing regulation make provisions for carry-over and storage aid for certain products.

The aims of all the intervention arrangements and indeed the whole marketing regulations are to ensure support for, and if possible increase, fishermen's incomes and secure fish supplies at reasonable prices for consumers. Nevertheless, the marketing regulations and the pricing arrangements have failed to attain these objectives. There is an increase in the total income of E.C. fishermen in the twelve Member States of 4.2%, from 2256 million ECUs in 1983 to 2350 million ECUs in 1988 [85] while fishermen in the twelve Member States increased in number by 13% from 280,311 Community fishermen working full or part-time in 1983 to 317,404 in 1988. When the increase in fishermen's income is coupled with the increase in fishermen there is a loss in the average per capita income of fishermen of 8% between 1983 to 1988. This loss of income is emphasised more when it is compared with the total increase in the GDP in the Community of the twelve of 40.3% during these years [141]. At the same time the price index of fish in the twelve Member States rose at an average rate of 37% [148], while landings decreased by 11.2% [125,148]. Thus, although the income of fishermen declined, the consumer paid more for fish without any benefit to the producers. It could be suggested that the loss of fishermen's income per capita is the result of declining landings, coupled with an increase of the fishing workforce. However, it is not possible to conclude with certainty about the scenario which caused this decrease and the interrelation of landings with the number of fishermen, since there are no robust time-series data available. Nevertheless, one cannot fail to observe that the way of promoting fishermen's income by applying such a common marketing policy with its intervention scheme has proved ineffective. It can be said briefly here that the best way to promote fishermen's incomes is by decisive measures for restructuring the industry, by proper management of stocks and by promoting the demand for fish.

In addition to the overall deficiency of this regulation, the Court of Auditors' report [68] has found particular deficiencies in the practical implementation of the price arrangements, particularly with regard to autonomous withdrawal prices and to the carrying out of obligations under the marketing regulations by Member States authorities. These obligations concern the notification to the

Commission of national rules and price arrangements. In a recent report of EUROPECHE [87], the carryover mechanism is alleged to suffer from a cumbersome administrative system, characterised by obscurities arising from the lack of a clear-cut separation from the withdrawal system and the lack of a strict control over the ownership of products, once they are considered unsold.

Under the new regulation autonomous withdrawal price arrangements are bound to be temporary with the consequent abolition of the autonomous regionalized price mechanism in the future. Despite the deficiencies in the practical implementation of this system reported by the Court of Auditors, this system is in principle fair. It is unquestionably true that there is a difference in consumers' tastes for fish products throughout the Community. Specifically, among the fish included in Annex VI of the 1992 marketing regulation [72] and subject to the autonomous withdrawal mechanism is horse mackerel. For this fish, there is a great variation in consumer tastes in the Community (Figure 6.8). For example, the price per tonne of horse mackerel in Portugal, where there is a strong market for human consumption for this fish, is nine times higher than the price per tonne in the U.K. market where there is almost no demand for horse mackerel for human consumption. As the quantity of horse mackerel supplied to Portuguese ports (23293 tonnes) is six times more than to U.K. ports (3774 tonnes), this price formation cannot be attributed to supply but is mainly due to diverse consumer tastes. Therefore, the future substitute for the autonomous withdrawal price must be designed to take into consideration the wide differences in consumer tastes and to adopt more strict measures for national authorities to keep detailed stock records for the species concerned and to communicate with the Commission on time. However, the effectiveness of the intervention scheme of the E.C. cannot be looked at in isolation before considering trade policy for reasons that will become apparent below.

The Community's trade policy

Intra-Community trade

The new 1992 marketing regulation as well as its predecessor in 1981 contains no provisions for dealing with trade in fishery products between Member States. In this case article 38(2) of the Rome Treaty comes into effect, for it states that where a regulation laying down a Common Organization of the Market for a particular sector of agriculture contains no rules relating to intra-Community trade, the general provisions of the EEC Treaty governing trade

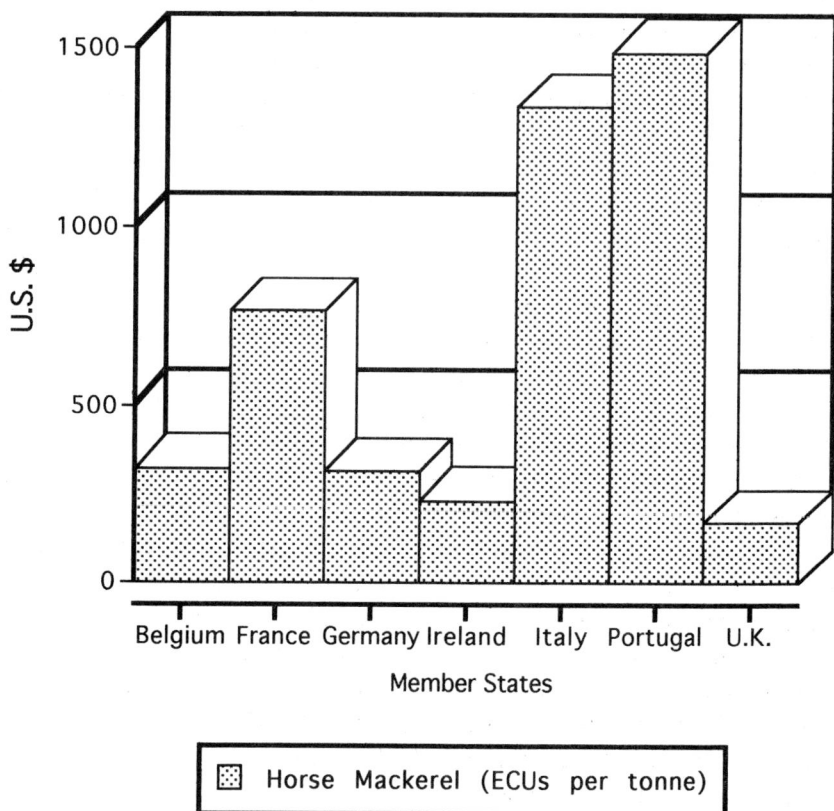

Value, in U.S. $, by national landings in domestic ports of seven Member States, per tonne, for Horse Mackerel in 1990

Figure 6.8 Price per tonne of horse mackerel in seven E.C. states

source: Review of fisheries in OECD Member Countries, Paris 1993

between Member States apply. These provisions can be found in articles 9-17 and 30-37 of the Rome Treaty. The products to which the provisions apply are, according to article 9(2) of the Rome Treaty, "products originating in Member States and products coming from third countries which are considered to be in free circulation in Member States". According to article 10(1) these are goods for which "import formalities have been complied with and any customs duties or charges having equivalent effect which are payable have been levied in that Member State, and they have not benefited from a total or partial drawback of such duties or charges". According to a 1968 regulation [88] fishery products originating in Member States are considered to be any fish caught by a vessel registered or recorded in a Member State and flying its flag.

The rules governing the trade in the above fishery products are defined in articles 13-16, 30 and 32-34 of the Rome Treaty, in articles 31-38 and 42 of the 1972 Act of Accession and in articles 35 and 64 of the 1979 Greek Act of Accession. According to these articles, all customs duties on fisheries products, any pecuniary charges imposed unilaterally on domestic or foreign goods, even though they may not be customs duties in the strict sense (89), and furthermore all quantitative restrictions and measures have to be abolished for intra-Community trade and cannot be re-introduced (articles 12 and 31 of the Rome Treaty). In the case of the two new Member States, Spain and Portugal, according to the 1985 Act of Accession [43] all custom duties were to be abolished by 1 January 1993. An exception is made for sardines for Spain and for tuna for Portugal where customs duties are to be abolished in 1997. If during this period imports of sardines or tuna lead to a disturbance of the market, minimum import prices may be imposed (articles 170, 173, 357, 360 and 362). In addition, article 95 of the Rome Treaty provides that Member States cannot impose any internal taxation on the product of any other Member State in order to circumvent the prohibition of customs duties to offer protection to other products. If a Member State introduces a tax on imports in such a way as not to infringe the above ruling, the Member State from which the goods are exported may reimburse its exporters for the import tax levied. This reimbursement according to article 96 of the Rome Treaty must not exceed the amount of any tax paid.

Although the provisions of the E.C. for free intra-Community trade are clear cut in theory, there have been some cases where Member States have restricted imports from other Member States by one way or another and which have been found by the European Court to act contrary to Community law. For example, in 1980 Belgium and in 1982 France required that imported fish be subject to health controls long before the application of common health standards in the E.C. (see above and [63]). The European Court in the Belgian case [90] ruled that such controls had an equivalent effect to quantitative restrictions. For France, the case was settled without being referred to the Court [91]. Another example is that of the U.K., which restricted access to Irish fishermen to a number of landing ports on the grounds that more effective control of conservation measures could take place. In this case the Court accepted the national legislation [93]. Italy in 1981 put a series of obstacles in the way of Danish imports, and the Commission was unable to agree whether the obstacles were contrary to Community law [94]. After the application of the Single European Act in July 1987 towards the future "blue Europe", intra-Community trade in fishery products has to become, in theory, a strong unilateral front for the Community to trade with third countries in order to reduce its structural deficit in fish which is increasing substantially from year to year. Nevertheless, as becomes evident in the analysis below (6.5.3.), the lack of a genuine common market in fisheries products is likely to persist for long.

Trade with third countries

Import Regulation Throughout the history of the European Commmnity, imports provided a considerable source of supply in fishery products (Figure 6.4, page 160), since the Community is far from being self-sufficient. Given this dependence on imports the E.C. regulators attempt to protect producer's incomes against extra-European competition. The legal rules governing such trade consist of custom duties, the establishment of reference and free-at-frontier prices and export refunds, all of which will be discussed in the following sections.

1 *Common Customs Tariff (CCT)* The CCT aims to ensure that the principle of Community preference is applied. The CCT has been strengthened under GATT (the General Agreement on Tariffs and Trade) since the early days of the Community (The Dillon Round, 1960-61) and took its first consolidated form in the 1968 regulation [88] after the Kennedy Round 1964-67. It applies to several fish products. During subsequent GATT negotiations (The Tokyo Round 1973-79) further changes and concessions have been made for third countries. Since then the CCT has been frequently revised. An amended version of the fishery products and the range of duties that are applied are published in an Annex by the Community each year. The consolidation process of the CCT provides the necessary tariff protection in relation to situations which currently exist in developed countries with a market economy. According to data, revealed by EUROSTAT-COMEXT, three quarters of the Community's imports (78.65%) of fishery products (by value i.e. 5000 million

ECUs) in 1991 take place under the rules of GATT. Nevertheless, this tariff contains a large number of exceptions with almost two thirds of Community imports covered by exceptional arrangements [104]. Furthermore according to article 20 of the 1992 marketing regulation [72] in emergency situations and in supply difficulties or under international trade commitments, the Council may decide to suspend the CCT wholly or partly.

The current rates in the consolidated tariff and the fishery products concerned are listed under Annex VII of the 1992 regulation. In general the tariff is up to 75% for processed products and the range varies from 10% to 18% for most fresh, chilled, and frozen products. Some commercially important fishery products have a much lower level of protection, for example salmon has a 7% tariff, cephalopods 6 to 8% and there are a number of zero duty tariff products e.g. salt cod, herring, frozen cod fillets, tuna.

The exceptions to the CCT take the form of a reduction in the tariff. They result from international undertakings or from unilateral decisions by the E.C.. Their scope is usually beyond that of the CFP and is concerned with a broader frame of a general policy on trade and development. These are as follows:

I *Exceptions by Treaty* These exceptions are agreed between the Community and a group of third countries. Under the Lomé III Convention [96] with the 66 or so African, Caribbean, and Pacific (ACP) countries, products originating in those countries have been granted access to Community markets completely free of custom duties, without reciprocal concessions.

Other bilateral treaties with tariff exception provisions are those that aim at the development of Mediterranean countries such as Egypt [97], Algeria [98], Tunisia [99], Morocco [100] and Turkey [101]. In addition a series of annual regulations were applied for Malta, which granted duty-free access for her products exported to the Community market. Moreover, commercial agreements with the European Free Trade Association (EFTA) allow specific reductions for certain products.

II *Exceptions under the Generalised System of Preferences (GSP)* The GSP is applied unilaterally by the E.C. to developing countries without reciprocal concession. The GSP arose at a meeting of the U.N. Conference on Trade and Development (UNCTAD) held in New Delhi in 1968. The countries eligible for GSP are divided into two groups according to their level of economic development depending on their inhabitants' level of income. The first group consists of "developing countries" that benefit from tariff reductions for certain products, and the second group termed as "the least-developed countries" benefit from complete suspension of custom duties for certain products including fishery products. The treatment of "the least developed countries" under the GSP is also applied by the Community to Equador, Peru, Colombia, and Venezuela to support them in their campaign against drugs. In addition, the GSP is applied to Czechoslovakia, Hungary, and Poland, though for Poland it does not cover fishery products as negotiations have not finished yet.

III *Tariff Reductions* According to a Commission report in 1987 [127] the supply policy of the E.C. for fishery products has been established with the additional consideration of guaranteeing supplies to the processing industry and allowing the E.C. processing sector to remain competitive without endangering the balance of the community fleet.

In order to fulfil this aim the Community unilaterally allows a number of imports to meet the specific needs of the fisheries sector at a reduced level of duty. Since 1986 the Council in accordance with article 28 of the Rome Treaty produces regulations about tariff reductions, some of which do not involve any quantitative restrictions and are classified as suspensions, while others involve such restrictions and are classified as tariff quotas. These tariff reductions fall into three main categories of fishery products:

1)   Unprocessed raw whole fresh or frozen white fish
2)   Semi-processed products which represent an intermediate raw material for the processing industry. These are mainly blocks of frozen fillets of hake, pollack or saithe.
3)   Salted cod-based products. The tariff protection scheme for these products was created especially for the southern Member States where there is a strong tradition of consuming salted products and the Community is unable to satisfy the demand. Tariff protection of these products was not applied before 1985, the year of the Spanish and Portuguese accession.

2 *Reference and Free-at-Frontier Prices* In addition to the system of customs duties, in order to preclude disturbances to the intervention mechanism of the Community (described above) and to give Community fishermen adequate protection from adverse competition from imports, the Community introduced, in article 22 of the 1992 marketing regulation, a system of reference and free-at-frontier prices for fishery products imported from third countries.

Reference prices are to be fixed each year for products specified in Annexes I,II,III,IV (B) and V of the 1992 marketing regulation [72]. The reference price is established according to the fishery product that is covered. For example, for live fresh, or chilled, fishery products of Annex IA and ID the reference price is equal to the withdrawal price; for sole, edible crabs, and Norway lobsters the reference price is to be equal to the Community selling price, and for the frozen products of Annex II the reference price is derived from the guide price according to specific calculations. The reference price for tuna products for use by the canning industry is specified in Annex III and is equal to the weighted average of the free-at-frontier prices (see below) during the three immediately preceding years.

For the same fishery products to which the reference price is applied, a free-at-frontier price is established. This is based on the prices recorded by Member States for the usual commercial quantities which are imported into the Community at representative ports and which are reduced:

i) by an amount equal to the customs duty in the CCT,
ii) by any charges levied on these products and
iii) by the cost of unloading and transport from Community frontier crossing points to those markets or ports (according to article 22(3) of the 1992 marketing regulation).

According to the above regulation, when a free-at-frontier price for a given product imported from a third country is lower than the reference price or when significant quantities of the product are imported, then the corrective actions listed below may take place:

1) The re-imposition of the CCT duties.
2) For products listed in annexes I (A,C,D,E), II, IV (B), and V the free-at-frontier price may be required to equal the reference price.
3) Where import prices for megrim, herring and tuna species are lower than the reference price, imports may be subject to a countervailing charge which must be in accordance with GATT rules. This charge is to be equal to the difference between the reference price and the free-at-frontier price and it is added to the customs duties in force.

Which combination of the above measures is to be taken is for the Management Committee for Fishery Products to decide. However, in the intervals between the periodic meetings of the above Committee, the Commission can take action.

3 *Particular provisions for certain products* A special system of reference prices for carp, trout and salmon operates under article 23 of the 1992 marketing regulation that aims to obviate "disturbances caused by imports at abnormal low prices". The reference prices for these fish are to be fixed on the basis of the average prices recorded during the preceding three years. This special system has been laid down for carp since 1974 [102]. Then in the 1981 marketing

regulation [58] trout were introduced under this scheme and recently the 1992 regulation [72] included salmon. If the free-at-frontier price is lower than the reference price, imports of these products are subject to a countervailing charge, which is calculated in a similar way to that used in the general system of reference and free-at-frontier prices, described above.

Another special measure that aims to control imports is applied for sardines and tuna species under article 21 of the 1992 new marketing regulation. For a period of four years the annual imports of these species are limited to a maximum equal to the total volume of imports recorded during 1991. Each year there is a permitted annual rate of increase that is based on the 1991 "reference year" amount of imports multiplied by an arithmetical average derived from the rates of development of consumption for these products in the Community during the two years preceding the reference year [103].

Critique of the Community's trade policy

Any measure, not only those that aim at the Common Organization of Market for fisheries but virtually for any sector of fisheries must be based on both the demand and the supply of these products. In recent years, the changes, which have taken place in developed societies, in the lifestyle at home and at work as well as the persistent advice of nutritionists have altered eating habits and produced new dietary requirements. In recent years there is a gradual increase in the involvement of the processing industry which produces more value added fishery products. The Community now possesses an industrial and trading complex covering a range of fishing activities, from catches through the processing of products at sea or on shore (canned, smoked etc.). However, this system is heavily dependent on imports from third countries (Figure 6.9) with as a consequence the lack of a self-sufficient internal market in fisheries. Given also the state of resources, the completion of the single market is unlikely to bring a reduction in imports from outside the Community. Figure 6.9 indicates the flow of Community trade for fishery products by value over the years. This trade shows a structural deficit which is increasing by some 15% to 20% per year. According to data from OECD published in 1993 [125], the Community imports of 8986 million US$ for all fishery products make it the world's second largest importer after Japan (10,033 million US$).

In terms of volume, in 1992 intra-Community imports consisted of 67.7% of fresh or frozen fish, crustaceans, molluscs and other aquatic invertebrates and 12.69% of processed fish products in a total of 2,458 thousand tonnes [126]. The extra-Community imports consisted of 55% fresh or frozen fish, molluscs and crustaceans and 12% processed fish products in a total of 4,096 thousand

million ECU

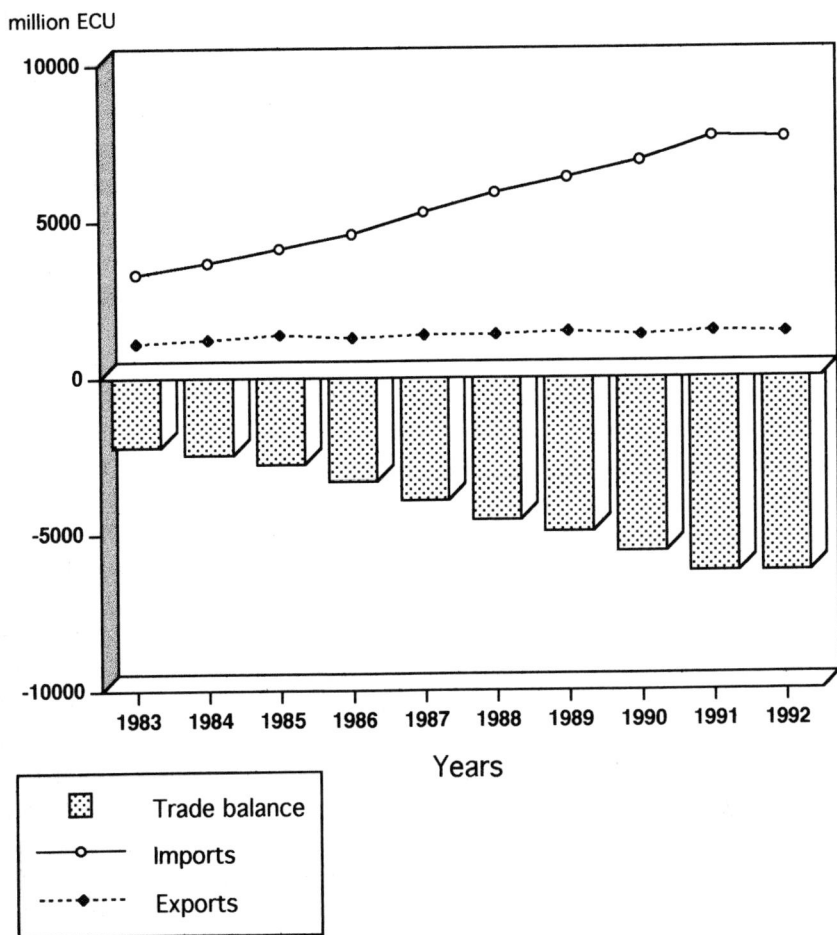

note: Data1983-1985 for 10 Member States, data 1985-1992 for 12 Member States

Figure 6.9   Flow of Community trade in fishery products during 1983-1992

source: D.G. Fisheries, XIV-A-3, situation at 30.VI.1993

183

tonnes [126]. As far as price per tonne is concerned, the intra-Community trade of processed fishery products is 18% higher compared with the extra-Community imports of processed fishery products and this is probably due to higher costs of the processing industry in the Community [126]. It should be noted that the extra-Community imports consist mainly of unprocessed luxury products with high added value when processed in the Community, i.e. tuna, salmon, crustaceans, molluscs and white fish for processing, i.e. cod and hake. These products are also at the root of the Community's external trade deficit, for example one half of the entire deficit for 1989 was due to shrimps, 20%, cod, 13%, and salmon, 11% [126].

However, given that the Community is not self-sufficient and relies greatly on imports of fish products to satisfy its demand, the liberalisation of the trading regime with third countries, established under GATT, is beneficial to the consumer and the processing industry of the E.C.. Under GATT rules, imports of fishery products move to the gradual dismantling of the Community's customs barriers which nowadays cover one quarter of the Community's imports of fish products. This situation is beneficial to the Community's processing industry and consumer since both can obtain fish products (most of which are luxury products) at lower price than if there were a tariff added to the price.

However, this situation is not beneficial to producers who have to compete against cheap imports from third countries. The Community aims "to protect" its fishermen from harmful import competition via the establishment of reference and free-at-frontier-prices and other measures that are mentioned above. The net result of this system of subsidies is to underline the inefficiency in the Community's fishing industry in comparison to others who can afford to sell cheaper fish products.

Although this Community system aims to secure producers' income it has failed many times since 1976, when it was first introduced. It failed completely to offer any protection to fishermen when crises arose in the late 1970s and the early 1980s and during the recent crisis of early 1993, when the market in fish was flooded by low-priced and often subsidised imports from third countries. The particular failure of the system can be attributed to many reasons that all lead to the conclusion that such a system was bound to fail in the first place. The main problems of this scheme is that reference prices are set at such a low level that they do not activate the mechanisms "to protect" producers in crisis situations. Another point is that for reference prices and the other measures to offer protection, the constant observance of import prices is required. This may work for small fluctuations of import prices. However, even if this can be achieved it is impossible to protect producers in gluts of low-priced imports, as

184

happened on the above two occasions. Another problem, heavily emphasised by Europeche [87], is that the system is not flexible, is not simple and functions cumbersomely at the procedural level.

This procedure that the Community follows for any problem arising in the market for fish with regard to the withdrawal arrangements and the pricing mechanisms has to be initiated by the chairman of the Management Committee for Fisheries Products, who is also a representative of the Commission, either on his own initiative or at the request of the representative of a Member State on this Committee. The chairman submits to the Committee a draft of the measures and the Committee gives its opinion on the draft adopted in accordance with the qualified majority voting procedure which applies in the Council (article 148(2) of the Rome Treaty). The chairman does not vote. If the Committee approves the Commission's draft then the Commission adopts the measures which apply immediately. If the Committee does not approve the draft, the Council may within a month modify the Commission's measures. This time-consuming procedure adds to the criticism that the system is not efficient when it comes to remedy crises such the above. The only drastic way that the system envisages a remedy to such a crisis is by the suspension of imports. Although the procedure for such urgent action diverges from the Management Committee procedure and the Council has a month to adopt the Commission's proposed measures by a majority of votes, it uses up precious time before a Council decision is made.

With regard to the consumer, this Community trade policy is totally senseless. It seems that the Community considers that the benefit that the consumer obtains by the gradual abolition of custom tariffs is too good to be true and thus obliges its consumers to be indirectly taxed in order to support this system for the benefit of the nation that exports the fish! In other words, even when the consumer and the processing industry could obtain imported fish products at lower prices they have to be taxed by the Community to a level at least equal to the reference price to maintain the senseless trade system. The final benefit goes to the exporters, who although willing to receive less, know that there is no sense in undercutting the reference price in order to sell their products. They, therefore, appropriate the difference between the reference price and their supply price.

Thus, the trade policy of the Community should not be based on these measures, which although designed to protect fishermen's incomes have not fulfilled their aim and also produce negative effects for consumers and the processing industry, denied the benefits of cheap imports. A more sensible policy, given the state of stocks, the Community's reliance on imports and the dismantling of customs duties would be through a more careful consideration of the principle of market access in return for access to resources. One way that the Community can protect fishermen's income and keep the fishing fleets active is not by subsidies but by exerting more effort to use access to its market as a means of bargaining for fishing rights in third countries' waters. This principle of access to markets in return for access to resources is applied in the Community's fisheries relations with third countries and it will be discussed below.

Access to third country waters. Bilateral agreements

The Community's arrangements with third countries for access to resources constitute an essential element of the CFP. It is significant that in the Community budget of 1993 fisheries agreements represented more than 25% of the total expenditure (Figure 5.2 page 119). This means that it is considered to be the second most important issue after the structural policy.

These arrangements with third countries aim to ameliorate the detrimental economic and social consequences for Community fishermen which may have arisen from the extension of fisheries zones. As will be discussed further below the Community fleet has surplus capacity in E.C. waters and bilateral agreements appear as one of the main elements to redirect and control fleet capacity by creating fishing opportunities in waters under the jurisdiction of other coastal states or in international waters covered by international conventions.

For example, according to a "Research and Documentation paper" produced by the European Parliament in 1991 [104], 16 fisheries agreements between the E.C. and ACP (African, Carribean and Pacific) countries allow for fishing rights to the Community fishing industry which employs fishing vessels of a total of 135 thousand GRT for all types of fishing activities (i.e. around 300 Community vessels). In general, 4,000 E.C. vessels are authorised to fish in the waters of third countries.

According to negotiating guidelines under article 43 of the Rome Treaty these agreements must be adopted by the Council with the prior consultation of the European Parliament. This consultation, although mostly purely formal, is usually concluded by the Subcommittees of the European Parliament on Fisheries, on Development and on Budgets. In practice, Parliament's opinion is sought after the negotiations have been concluded by the Commission and sometimes even after the ratification by the Council (e.g. in the case of the E.C.-Greenland agreement, see below). This practice is totally unsatisfactory if one considers the place of these agreements in the Community fisheries budget in years 1983-1993. In 1990, the year of the second agreement between the E.C. and Greenland such outlays reached a maximum (Figure 6.10) that represented 39% of the total payment appropriations for the CFP (446 million ECU) [105].

The bilateral agreements are so called "framework agreements" and establish general conditions for fisheries relations between the two contracting parties. These include general provisions on access to fishing zones, compensation or licence fees, where appropriate scientific co-operation in the region involved, technical support where appropriate and procedures for the settlement of disputes. In every case the balancing of mutual interests is based on the available information, both scientific and economic, deriving from the main international organizations (FAO, etc.), from the third country in question or from experimental surveys partly financed by the Community.

All bilateral fisheries agreements involving the Community and third countries must be Community agreements according to the principles of the CFP mentioned before. This principle has also applied to cover the integration of the former GDR into the Federal Republic of Germany. However, there are certain exceptions to this principle. For example, France concluded fisheries agreements with Canada in March 1989, with Korea in April 1991 and with the USA in February 1991. In addition Denmark made agreements with Lithuania in 1991 and with Iceland, Canada and Norway in March 1991. Most of these exceptional agreements concern certain fishing areas and certain fishery products [120].

Following the accession of Spain and Portugal all their previous agreements with third countries have been managed by the Community. For example, the Spanish and Portuguese bilateral agreements with South Africa were due to expire for Portugal in 1989. After Community negotiations with South Africa it expired in March 1992 which was also the date of expiry of the Spanish agreement. In total there are currently 23 agreements which the Community has signed or is about to sign with third countries. There is a series of negotiations that have been started by the Council which has authorised the Commission to start them (Table 6.1). Some negotiations started as far back as in 1987 but they have not been concluded yet, for example negotiations with Gabon, Ghana and Liberia. Other new agreements that are about to be approved by the European Parliament and the Council (November 1992) have been made with the newly recognised states of Estonia, Latvia and Lithuania [106].

Most of the bilateral agreements with third countries are those based on financial compensation, as happens for trawlers and for tuna fishing vessels (Table 6.1). However, the most costly agreements to the Community are these of mixed type which will be discussed below. A complete list of the

187

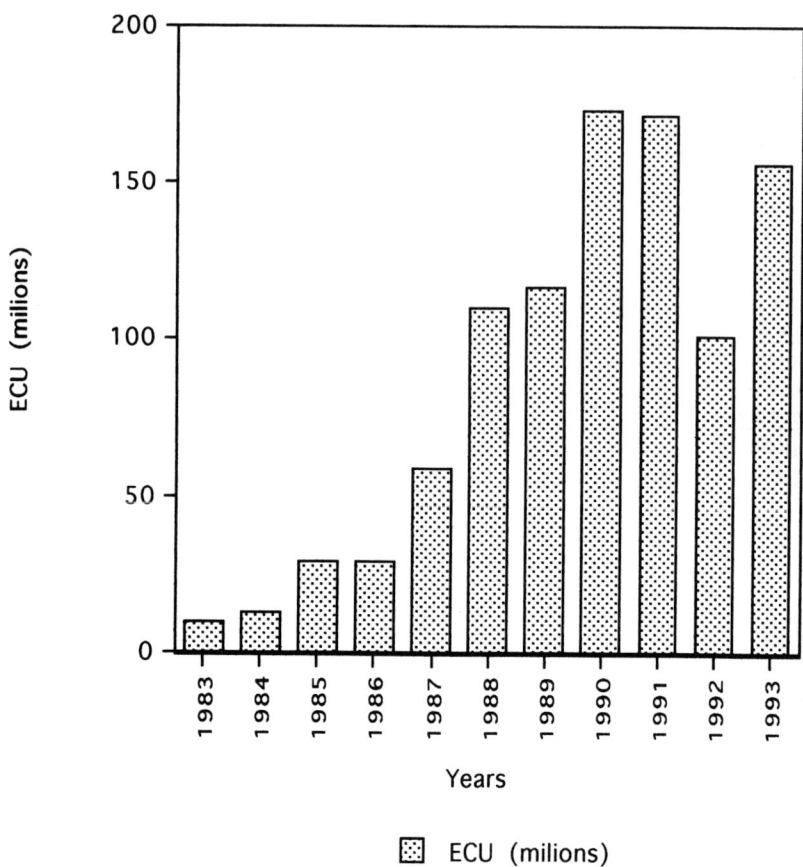

Figure 6.10 Evolution of E.C. expenditure on fisheries agreements with third countries

sources: Yearly Official Journals on Final adoption of the general budget of the European Communities for the years 1983-1993

Table 6.1 The Community's bilateral fisheries agreements with third countries.The situation at 1 January 1993

Type of agreement	Third countries	Annual amount of financial compensation from E.C. budget (ECU million)
Reciprocal agreements	Norway Sweden Faeroes Finland	0.765
Agreement based on access to resources and markets	Canada	
Agreements based on access to surplus resources	USA	
Agreements based on financial compensation	São Tomé and Príncipe Gambia Angola Mauritania Comoros Islands Mauritius Sierra Leone Cape-Verde Islands Madagascar Mozambique Ivory Coast Senegal Tanzania Seychelles Gabon	0.725 1.385 9.25 9.670 0.485 0.650 2.825 0.870 1.266 0.530 2.370 15.000 0.560 3.300 3.050
Mixed type agreements	Morocco Greenland	70.375 34.250
New agreements not yet approved by the Parliament and the Council	Estonia Latvia Lithuania Morocco	
Negotiations or intended negotiations with third countries	Comoros, Kenya, Somalia, Djibouti, Ghana, Liberia Sri Lanka, the Maldives, Nigeria, Namibia, Iceland	
TOTAL of the 1993 E.C. budget on fisheries agreements		156.561

sources: Commission report SEC(91) 2288 final, Doc. PE 140.600, E.C. budget for 1993 O.J. (8.2.1993)Ll31/1446-1452 and for each agreement in the relevant O.J.

Community's fishing agreements, still in operation in 1993, together with their particular and total financial costs, as they appear in the general budget of the Community [107], is presented in table 6.1. They are normally concluded for several years and include a renewal provision which many times is tacit.

The various bilateral agreements with third countries can be divided in the following five categories.

Reciprocal agreements

These have been signed with the Scandinavian countries (Norway, Sweden and the Faeroe Islands) which have common borders with the Community, and with Finland. The agreements involve access on a reciprocal basis to a specific part of the stocks in the fishing zone of each country with no financial implications. Each year, since 1981, bilateral or trilateral meetings have been held to establish fishing rights for the following year in the light of the scientific assessment of stocks and to seek to harmonise technical measures. Then the twelve Member States divide the catch quotas by negotiating with the Commission according to formulas based on a reference index established in 1983. In practice the above scheme is implemented through co-operation within ICES.

According to the reciprocal agreements with the above third countries, the Community has access to about 239 thousand tonnes of fish from Norway, 48 thousand tonnes from the Faeroe Islands and 10 thousand tonnes from Sweden each year. Since 1985 the year of Spanish and Portuguese accession the Community has received a special quota allocation of red fish from Norway that varies each year between 1000-1500 tonnes and which is not shared according to the reference allocation of 1983. This quota is granted exclusively to Spain and Portugal in order to facilitate their accession to the Community and it represents the quantities fished in these waters by Spain and Portugal before their accession. Furthermore, among the Member States that benefit from the agreements with Norway, the most important partners for this type of agreements in terms of quotas allocated in the North-East Atlantic and in the Barents Sea, are Denmark, the UK, Germany, and France. As for Sweden the Member States that have access to Swedish stocks are Germany and Denmark. Since 1980 [121] the Community has been making Sweden an annual payment of 765 thousand ECUs towards the costs of salmon bred and released in the Baltic Sea (Table 6.1).

In general the scheme of reciprocal agreements operates to the satisfaction of all members concerned. The most important reason for this is the availability of the considerable amount of scientific data that ICES provides, the relatively adequate knowledge of the North Sea and neighbouring waters fish stock levels and the co-operation of experts from all the countries under the aegis of ICES. The problems that sometimes arise are due to negotiating difficulties in allocating the Community quota between the Member States and the harmonisation, as far as possible, of technical rules that apply to different fisheries. For example in the Skagerrak and Kattegatt and in the Baltic Sea the countries concerned have agreed on technical measures. In the North Sea the technical measures described above for the Community fishermen are not the same as those for fishing in Norwegian waters. Usually Community fishermen use different mesh sizes when they are fishing in Norwegian waters.

In the case of the Faeroe Islands where the 46 thousand inhabitants generate about 98% of their GNP from fisheries and about 95% of their exports earnings [128], the Community has taken this into consideration. The reciprocal agreement with these islands is to the voluntary disadvantage of the Community. The two parties have had difficulties under the aegis of NASCO (North Atlantic Salmon Conservation Organization) in reaching an agreement on salmon fishing in the Faeroe Islands.

The agreement between the Community and the USA

This agreement started in 1977 and since then it has covered the available surplus of fish stocks in the American fishing zone. Under this agreement Community fishermen with US involvement must set up joint ventures and buy certain quantities of fish ship-to-ship from American fishermen. The quantities derived from direct fishing or from purchases from American fishermen or from purchases of processed products are defined annually. Before 1986 there were four Member States with fishing rights in American waters, namely Italy, Portugal, Spain and Germany; since then only the Netherlands has made use of this right. The Community reacted and sought compensation for the damage done to its fleet but the matter has not been settled yet. In February and March 1991 the two parties exchanged letters in order to extend their agreement until 1993. The Community accepted under pressure from the US government a UN resolution [108] on the banning of long drift nets and on conserving pollack in the Bering Sea. Finally, a decision was taken in favour of the continuation of the agreement by the Council in 1991 [109] without seeking the opinion of the European Parliament.

This agreement was concluded in December 1981 for a period of six years [112] with a provision of tacit renewal. Under this agreement the Community was granted access to Canadian waters in exchange of tariff concessions for imports of certain Canadian fishery products into the Community market. Although the agreement was never revoked and is formally still in force, it has not been applied since the end of 1987.

The arrangement was never really satisfactory for Canada which protested to the E.C. in an exchange of letters that the tariff concessions were only on paper and in practice Community administrative restrictions never allowed a full application of the reductions in customs tariff. The situation became more acute after 1985 when Canada following the NAFO recommendations refused to issue licences to E.C. vessels for fishing cod in NAFO area IIIL. Then, the Community applied unilateral quotas for this area in 1987 in retaliation. The crisis deepened during 1990-1991 when the dialogue between Canada and the Community failed to reach an overall agreement and reached a climax in 1991 under the GATT Uruguay round. During this GATT round Canada backed by the USA insisted on discussing only commercial aspects, i.e. customs duty and tariff concessions, while the Community wished to discuss overall trade measures including access to stocks [110]. The crisis has non been resolved yet, although the Community as a gesture of goodwill proposed and adopted a series of measures and regulations [111] in favour of the Canadians. These regulations apply to Community fishermen whose fishing vessels operate in the areas concerned and lay down limits on catches, technical conservation measures (minimum mesh sizes, limits on catches and attachments to cod ends) and make provisions for the better processing and transmission of scientific and statistical data. In addition, to ensure that these measures are applied, all the Community vessels fishing in the North-West Atlantic have to accept on board observers approved by Canada, observers who have the duty to record and report upon the fishing activities of these vessels. Recently, in May 1993 [130] the Commission concluded an agreement with Canada which covers the whole range of issues affecting their fisheries relations. For example, issues of access by E.C. fishermen to Canadian ports and surplus resources in Canadian waters are included, as are also the opportunity for E.C. fishermen to conclude private contracts with Canadian fishermen who cannot use the quotas allocated to them.

The agreements between the E.C. and the ACP countries are formed by the combination of co-operation measures -development projects for fisheries- which fall under the Lomé IV Convention [129] (Article 68) and fisheries agreements which provide for financial compensation in exchange for fishing rights. In financial terms, the funds allocated to fisheries agreements are entered in the Community budget, whereas those used for development policy come from the Member States' contributions outside the budget. The principal species caught under these agreements are tuna, shrimps and high-quality demersal fish.

In institutional terms, the bilateral E.C.-ACP fisheries agreements are concluded for a period of two to three years, subject to tacit renewal. They contain an annex setting out the conditions for fishing and a protocol laying down the fishing rights, together with the financial compensation to be provided by the Community. In most of the agreements, a joint committee is established for the purpose of monitoring implementation. The supervision of the fishing activities is the responsibility of the coastal third countries.

The Community pays the financial compensation to the national treasury, with licence fees paid directly by the ship owners concerned in return for fishing rights. This financial compensation varies from one ACP state to another, depending on the catch possibilities and the commercial value of the products. However, the dominance of the "financial compensation" element is apparent since it represents an average of 33% of the total cost of the agreements currently in force (Table 6.1).

As far as the licence fees of the ship owners are concerned, these are fixed on the basis of vessel tonnage in the case of trawlers, or catch tonnage in the case of tuna fishing vessels. Additionally, the ship owner is obliged to conform with certain "development" provisions contained in the agreement, such as the compulsory landing of a proportion of the catches in the ports of the ACP country, local crew training etc.

The Community encourages the development of regional agreements between the ACP countries, as this would make resource conservation easier, especially for tuna-like fishes which are migratory. Examples of regional agreements are represented by the South Pacific Forum and the sub-regional committee on fisheries organised by a number of states in the Gulf of Guinea.

Unfortunately, the ACP countries are not very eager to embrace long-term development projects and they seem to prefer fisheries agreements. This is due to the fact that these agreements offer major short-term financial gains, which they can use without any constraints. In some cases the costs of the renewal agreements remain the same as for the previous. For example the recent agreement with the Seychelles in May 1993 [131] provides the same amount of compensation as did the previous one [132]. In other cases the cost of the agreements - compensation and licences fees - increases significantly on each renewal. For example, there was a substantial increase of 35% between the annual financial compensation paid to Mauritania under the first agreement in 1987 [133] and the second E.C.-Mauritania agreement in 1991 [134].

Mixed type agreements

These agreements are based both on financial compensation and commercial concessions to third countries in return for the Community's acquisition of fishing rights in third countries' waters. The E.C. has signed two agreements of this type, one with Morocco and another with Greenland.

The agreement with Morocco was signed in June 1988, was valid until 29 February 1992 [114] and amended in 26 March 1991 [115]. A new agreement with Morocco is soon to be signed based on the Commission's proposal to the Council [116]. This E.C. and Morocco agreement concerns almost entirely fishing off the Atlantic coast of Morocco. It is an important agreement for the protection of employment in the Community. For example, in 1988 more than 100 000 fishermen were dependent on the agreement and 736 Community vessels [117], mainly from Spain and Portugal (about 99 300 grt according to the 1988 agreement), gained access to fish in Moroccan waters. In return Morocco has the opportunity to export 17 500 tonnes of tinned sardines annually to the Community duty free.

Nevertheless, the financial compensation to Morocco is too high and this bilateral agreement costs the most to the Community. For the period of the four years the Community paid 272 million ECU in financial compensation, 6 million ECU for scientific programmes and 3.5 million ECU for study grants [114]. These include the costs for vocational training for local seamen and the use of local infrastructures as well as direct contribution to the setting up of joint ventures for the exploitation of the fishery resources. It costs an annual average of 70 million ECU which is twice as high as that of the E.C. and Greenland agreement, which has been criticised as one of the most costly, and also exceeds the total annual cost of all fisheries agreements between E.C. and the ACP countries (i.e. 60.28 million ECU) in 1991 [104]. However, the actual catches by Community vessels in Moroccan waters remain unknown.

The agreement with Greenland, after its decision to withdraw from the Community in 1985, was concluded in January 1985 [118] for a ten year period. The 1985 regulation and that of 1989 laid down the annual catches authorised for E.C. vessels in Greenland's waters and are presented in table 6.2. In return Greenland gained free customs duties and no quantitative restrictions for fisheries products and an annual compensation of 26.5 million ECU. The annual financial compensation increased with the last amendment of 1989 [118] by 19.2% and reached 34.25 million ECUs as table 6.2 shows. It is interesting to note that the above agreement caused considerable friction between the Community's institutions, since the Council failed to co-operate effectively with the European Parliament on the issue of the E.C.-Greenland agreement of 1989 [118]. In July 1990 [119] the Parliament expressed its deep dissatisfaction at being presented with a systematic "fait accompli" and made severe criticisms of this agreement as follows.

1   Spain and Portugal have been excluded from the quota allocated to Community although Spain had taken substantial catches in Greenland's waters during the reference period (1973-1978). After this agreement Spain and Portugal brought a case before the Court seeking the annulment of the 1989 regulation and the Council decision (118).

2   As table 6.2 indicates, the increase in the financial compensation by 29.2% was higher than the increase in quotas, 20.5%, which is totally unsatisfactory when one considers also the under-utilised quotas from the past agreement. Article 7 of the agreement also stipulates that authorised catches may be suspended without reducing the financial compensation. The European Parliament concludes that this agreement is more an agreement on financial aid to Greenland than a fisheries agreement based on a rational counter concession for authorisation to fish.

However, this criticism made by the European Parliament cannot stand. The increase in the real value of fishery products during the years 1985-1990 should additionally be considered. To decide if the European Parliament's allegation holds I used data published by the D.G. XIV [148] to calculate the average increase in value per tonne for extra-Community imports of all fishery products during the years 1985 to 1990 (i.e. the years of the first E.C.-Greenland agreement). This amounted to about 30%. Comparing the percentages given in the European Parliament's criticism with the 30% increase in the value of the extra-Community's imports for fishery products it is more reasonable to conclude that the Community's second agreement with Greenland (1990-1994) is in favour of the Community contrary to the European Parliament's opinion.

Because of the high cost of this agreement the European Parliament suggested that shipowners operating in Greenland waters should share the financial burden via licence fees.

Table 6.2   E.C.-Greenland fisheries agreement, authorized quotas and financial compensation

Species	A period 1985-1989 (tonnes)	B period 1990-1994 (tonnes)	Difference B - A (tonnes)
Cod	23,500	31,000	7,500
Redfish	63,320	52,320	-11,000
Greenland halibut	5,600	5,600	0
Halibut	200	200	0
Shrimps	4,350	4,350	0
Catfish	2,000	2,000	0
Blue whiting	30,000	30,000	0
Capelin	-	30,000	30,000
Total catches (tonnes)	128,970	155,470	26,500
Cost of the agreement (ECU)	26,500,000	34,250,000	7,750,000

source: Reg. 223/85 O.J. (1.2.1985) L29/8, Reg. 224/85 O.J. (1.2.1985) L29/13, amended by the Reg. 4054/89 and the Council Decision of 89/650/EEC in O.J. (30.12.1989) L389/65 and /80 respectively

Community membership of international fisheries conventions

All the international fisheries commissions tended to be more powerful before the introduction of the 200-mile zones in November 1976 since most fish stocks fall within this area. However, they continue to have an important role in fisheries management by regulating fishing beyond the 200-miles limit and by co-operating with coastal states in the management of stocks which migrate between these zones and the area beyond. Their main aims are to develop models for managing the most important stocks in each zone and to divide the fishing rights between the parties involved. For managing fish resources the main problems that these international organizations face is to keep a balance between their management plans and those of the coastal states and to impose technical measures for the conservation of fish stocks. For example, the resolution (No. 44/225) adopted by U.N. in December 1989 calls for a moratorium on the use of large-scale pelagic driftnets which reach or exceed 30 miles (48 Km) in length to catch living marine resources on the high seas of the world's oceans and seas [123]. This resolution was brought to the attention of the FAO regional organizations GFCM (General Fisheries Council for the Mediterranean) and ICCAT (International Commission for the Conservation of Atlantic Tunas) (Table 6.3) which produced a report in 1991 [124] that urges the need of taking such a measure.

There are cases where member states of such organizations can impose their will and offer part of their quotas arranged by international organizations to other contracting members, for example the agreement between Canada and the E.C. in the NAFO area. The division of fishing rights among the contracting parties is based on the historical fishing pattern when data from the past are available. Another problem faced by international organizations is the financing of monitoring and surveillance of fishing activities of the contracting parties. The Community is actively involved in an inspection plan in the NAFO area (as mentioned above).

Individual membership by Member States of international commissions has ceased and been replaced by a collective Community membership i.e. the Commission represents the Member States. This collective representation is endorsed by a European Court decision of 1976 [122] which was based on the Rome Treaty article 116. This article provides that Member States must "proceed within the framework of international organisations of an economic character only by common action".

Table 6.3 shows the various international organizations that are linked one way or another with the Community and the date when the Community became

Table 6.3 International Fisheries conventions linked with the E.C. and the date when the Community became a member where applicable

PART A.
- North-West Atlantic Fisheries Organization (**NAFO**); 1.1.1979
- North-East Atlantic Fisheries Commission (**NEAFC**); 17.3.1982
- International Commission for the South-East Atlantic Fisheries (**ICSEAF**); 12.1985
- International Baltic Sea Fishery Commission (**IBSFC**); 18.4.1984
- North Atlantic Salmon Conservation Organization (**NASCO**); 10.1983
- International Commission for the Conservation of Atlantic Tunas (**ICCAT**); 7.1984

PART B.
- Commission for the Conservation of Antarctic Marine Living Resources (**CCAMLR**); %.1980
- International Whaling Commission (**IWC**); 29.10.1979

PART C.
Regional international organizations set up by FAO.
- General Fisheries Council for the Mediterranean (**GFCM**)
- Fishery Committee for the Eastern Central Atlantic (**CECAF**)
- West Central Atlantic Fisheries Commission (**WECAFC**)
- Indian Ocean Fisheries Commission (**IOFC**)
- Inter-American Tropical Tuna Commission (**IATTC**)
- Baltic Sea Salmon Standing Committee (**BSSSC**)
- Shellfish Commission for the Skagerrak and Kattegat (**SCSK**)

PART D.
- European Convention on the Conservation of European Wildlife and Natural Habitats (**ECCEWNH**); 12.1982
- Convention on the Conservation of Migratory Species of Wild Animals (**CCMSWA**); 12.1982

PART E.
- International Council for the Exploration of the Sea (**ICES**); 1902
-.Convention on Conduct of Fishing Operations in the North Atlantic (**CCFONA**); 1967

198

a member. It is not the scope of this book to present a detailed analysis of the activities of each of them. The most prominent international organizations whose activities concern directly the CFP have been analysed in various places above and their operation and efficiency have been assessed. In the organizations in part A of table 6.3 the Community is a full member i.e. entitled to vote as a single body. For those in part B the Community is authorised to be a member while its Member States have the same status. This means that in the event that a Member State withdraws the Community will lose its right to participate. In addition, the Community is only entitled to vote when one of its Member States transfers its own vote to the Community. In Part C are those organizations under the aegis of the FAO where the Community gained its accession in November 1990. In part D are listed these organizations to which the Community has become a party and whose concern for fisheries is not their first objective. Part E contains the organizations which embody several Member States of the Community but not the Community as a whole.

7 The Common Structural Policy

The Common Structural Policy for the fishing industry was first established in 1970 by a regulation [135] that was based on the proposals of a Commission report in 1966 [136]. This regulation underlines the objectives of article 39 of the Rome Treaty (mentioned above) and aims according to article 5 to co-ordinate the structural policies of the Member States for their fishing industry. It laid down a number of obligations for Member States and for the Commission, designed to ensure such co-ordination. For example, it established, according to article 11, the Standing Committee for the Fishing Industry which required communication between the Commission and Member States on the structural situation of the fishing industry, an obligation that the Commission failed to comply with [137]. Furthermore, this regulation re-stated a 1964 regulation [138] which allowed the use of the Guidance Section of the European Guidance and Guarantee Fund (EAGGF) to promote the construction and modernization of coastal and pelagic fishing vessels and of the marketing and processing industry.

After the introduction of the 1983 CFP in January of the same year [27,28], the Council adopted a new structural policy on 4 October 1983 [143] which aimed to adapt the Community's fishing fleet to the catch possibilities inside the 200-mile EEZ by reducing the fishing capacity and to increase catch possibilities outside this zone. The instruments that the Community used to obtain this aim concern jointly:

1   A reduction of fishing capacity in Community waters, given the state of fish stocks, by scrapping and mothballing fishing vessels of certain categories.

2   A re-deployment of fishing capacity by encouraging shipowners to explore fishing possibilities, in particular outside the Community waters.

This policy provided aid for vessels over 24 metres in length to make exploratory fishing voyages in search of new fishing grounds and promoted co-operation with third countries for joint ventures especially in the Mediterranean or off the west coast of Africa.

3   A restructuring, modernization and development of the fishing industry by giving financial aid for the construction and modernization of vessels between 9 and 33 metres in length. In order to promote aquaculture the Community encourage programmes for constructing and modernizing aquaculture installations and the building of artificial reefs to restock certain areas, mainly in the Mediterranean.

To promote these structural measures the Community once more followed the familiar path of giving subsidies. At the same time as the 1983 regulations, a Council directive [144] set up a budgetary fund totalling 53.55 million ECUs to be used as financial assistance for these measures.

After the Spanish and Portuguese Accession in 1985 the Community was faced with an influx of fishing vessels that amounted to almost one third of all the other Member States' fishing fleet (80,942 motorised vessels). In an effort to deal with such an influx the Community gave priority to the classification and definition of the characteristics of fishing vessels. Thus, a Council Regulation on 22 September 1986 [145] aimed to define characteristics of fishing vessels such as length, breadth, tonnage and engine power by using identical rules, which would facilitate the legislative process and the application of the structural measures.

At the end of 1986 the Community adopted a new structural regulation [146] which fused together all the previous structural regulations and was designed (unsuccessfully) to remain in force for ten years from 1 January 1987. In addition to the usual new Community regulations to increase financial aid given for carrying out structural measures, this regulation also promoted fishing beyond the 200-mile zone by granting financial aid for building and modernizing vessels with a length in excess of 33 metres. This new aspect of regulation was in no way a decisive new measure to decrease and re-deploy the overcapacity of the Community fleet inside the 200-mile zone. It was an inevitable provision since the fishing fleets of Spain and Portugal, the two new Member States at that time, operated to a large extent outside Community waters. Under this regulation the increase in Community spending on the structural policy was from 250 million ECUs for 1983-1986 to 800 million ECUs for a five year period from 1986 to 1991. This implied a doubling of the yearly amount set by 1983 regulation from 83 to 160 million ECUs yearly under the 1986 regulation. Community aid, given to the fisheries sector through the structural policy must be accompanied by a national aid scheme in every Member State. This requirement gives rise to the possibility of unequal access to Community aid since available aid can be varied according to the national policy of each Member State.

Half-way through the period of existence envisaged by the Community for the 1986 regulation the Council at the end of 1990 adopted a new structural regulation [147]. This new regulation increased the structural funds to cover the structural requirements of the Spanish and Portuguese fleets. Moreover, it filled the gap in the 1986 regulation that contained no provisions for the "small-scale, non-industrial fishing" Community fleet (i.e. vessels less than 9 metres in length or less than 12 metres for those vessels engaged in trawling). The new elements of the 1990 regulation are as follows:

1   It reduced the financial aid for the construction of new vessels and the modernization of the existing fishing vessels. However, vessels employed in "small-scale, non-industrial fishing" became eligible for Community aid for construction and modernization as well as for scrapping.

2   It increased the aid for capacity withdrawal i.e. temporary laying up and permanent cessation of fishing activity through the scrapping of existing vessels.

3   It encouraged the reorientation of fishing activity by giving financial support to exploratory fishing by promoting redeploymernt operations and by the creation of joint enterprises.

This regulation came into force on 1 January 1991. In July 1991 the Commission adopted five more regulations and a Commission decision [148] which laid down detailed rules for the application of schemes for joint enterprises/ventures, for exploratory fishing and for re-deployment operations as these were envisaged by the 1990 structural regulation.

The Multiannual Guidance Programmes (MAGPs)

The MAGPs were first laid down by the 1983 [143] structural regulation. They were designed to be one of the main instruments for the implementation of the structural policy of the Community. The MAGPs were established in 1983 and were to run for three years up to 31 December 1986. They simply stated that Community fishing resources and opportunities were diminished and expressed the desire that the E.C. fishing fleet adapt to the availability of the resources.

The MAGPs were better defined in the 1986 structural regulation [146] (article 2(1)) They comprised a set of objectives applied to the 11 Member States for the reduction in tonnage and in engine power of their fishing fleets. The Community is determined that the terms of the MAGPs be strictly applied and financial aid is refused for national projects that are not in compliance with the aims of the MAGPs. In order for the Community to carry out the objective supervision of the MAGPs and to ensure that they are in compliance with the objectives set out in the Rome Treaty, Annex 1 of the 1986 regulation [146] aims to effect an evaluation of the relative importance of the fishing industry in the national and the regional economy of Member States. In addition, the Community seeks to take into consideration the situation of the fleet, by category of vessel and type of fishing, the region, the availability of the resource and the trend of prices in the market for fish products. The overall objectives of the MAGPs are to be achieved in stages. Throughout the whole period the Member States must inform the Commission about the progress in the capacity reduction of their fishing fleet.

MAGPs theory and practice, a critique

1983-1986 Fleet MAGPs The MAGPs of 1983 and the objectives for the reduction of the capacity of the fishing fleet set by the structural regulation 2908/83 [143] for a period of three years until 31.12.86 were in many cases established above the existing fishing capacity of certain Member States (Table 7.1).

It is evident from table 7.1 that the "objectives" of the 1983 structural regulations were unrealistic and represented real objectives only for Germany, Denmark and Netherlands. However, in the case of Germany the data given include fishing boats used for sport and therefore the actual capacity of the German fishing fleet was unknown. The great difference between the actual situation of the Greek fleet and the objective capacity is due to the fact that only the Greek inland fishing fleet appears in the data for 1982. Thus, the objectives of the 1983 programme were meaningless. According to the table 7.1 there was an overall reduction of 2.7% in the fishing tonnage capacity in the Community during 1983 to 1986. However, this reduction is due to only two Member States, the U.K. and Spain. In all the others there is an increase of the fishing fleet, whereas for Germany and Greece the situation is not clear because of unreliable data in 1982.

Moreover, the accuracy of the information that was given to the Community by the Member States at that time, the basis for establishing its objectives for the reduction in fishing capacity, is debatable. Although the Community

Table 7.1 1983-1986 MAGPs, E.C. fishing fleet capacity in Gross Registered Tonnes (GRT)

	Situation in 1982	Objectives of the 2908/83 regulation	Situation at the end of 1986
Belgium	21,845	22,000	22,846
Germany	86,106	78,479	59,000
Denmark	126,376	123,091	133,477
France	180,220	192,807	205,016
Greece	39,467	134,659	181,109
Ireland	39,448	45,300	49,000
Italy	262,270	275,255	284,631
Netherlands	106,370	66,800	126,632
U.K.	186,692	198,977	150,160
Spain	632,528	668,668	663,158
Portugal	184,265	215,930	194,971
Community	1,865,232	1,951,179	1,815,224

sources: Court of Auditors, Special report No 3/93 and D.G. XIV, XIV-A-3 30.6.1993

spending for construction, modernization of the fishing fleet and for aquaculture was increased considerably, from 118 million ECUs in years 1978-1982 to 156 million ECUs in years 1983-1986 and despite the establishment of the MAGPs that aimed to act as a prerequisite for granting financial aid for the restructuring of the fishing industry, the Community did not give appropriate consideration to the validity of the data about the fishing fleet capacity, that was given by the Member States. The situation would have been better if before the introduction of the MAGPs the Community had spent part of its financial resources to establish and maintain a Community register of fishing vessels. Community finance for this system could cover administrative expenses as well as the costs incurred by Community officials in scrutinizing the kind of information given by Member States. Although MAGPs had been running since 1983, it was not until 25 January 1989 that the Community adopted a Community register of fishing vessels [150] with a view to avoiding reliance on unchecked information given by Member States about their fishing fleet capacity. If the Community had established a Community register for fishing vessels before the application of the MAGPs, then the existing overcapacity in the fishing industry could have been tackled in a more efficient way and the financing of structural projects would be according to realistic objectives to reduce overcapacity.

In spite of the deficiencies of the 1983-1986 MAGPs, the 1983 structural regulation was the first that directed the Community fishing fleet to adapt to the available resources. Fortunately, after this regulation the Community was committed to follow this objective.

1987-1991 Fleet MAGPs The 1987-1991 MAGPs were initially set up by the 1986 structural regulation [146] and a final target for 31.12.1991 was fixed, based on an overall reduction of 3% in tonnage of fishing capacity in terms of GRT, and a 2% reduction in engine capacity in terms of KW, compared to the objective which had to be achieved by the end of 1986 under the previous programme. These objectives had to be met in successive stages by the end of 1991 with a rate that varied between Member States. Since the implementation of the 1987-1991 MAGPs the Community has adopted a series of 11 decisions, which were all amended, declaring that in such a way the MAGPs are "adjusted to reality". In fact, the Community realised that it is not practical only to set objectives to reduce fishing capacity if these are not to be followed by a variety of accompanying measures. The effectiveness of the series of the decisions and their amendments, which aimed to contribute to the fulfilment of the objectives, is dubious since they were adopted after the setting up of the objectives and tried hastily to amend previously unforeseen situations. Moreover, there were decisions taken only a few months before the end of the 1987-1991 MAGPs, as happened for the U.K. and Spain when a decision concerning the final objectives of the 1987-1991 MAGPs was adopted four months before the end of the programmes. They inevitably were to have little effect on the overall progress towards the targets.

A case, which was not covered by the objectives of the 1987-1991 MAGPs, was the shipbuilding projects which were carried out at that time authorised by the 1983 structural regulation [143]. As a result, contrary to Community expectations there was an increase of fishing capacity at the end of 1987 in the fishing fleet of eight Member States where data are available [152]. For these Member States there is an overall increase in fishing capacity of 4.2% from 1986 to 1987 measured in GRT. This increase persisted during 1988 compared to the fishing capacity of 1987 in Belgium (5.2%), Netherlands (27.7%) and in the UK. (6.2%) and was due to the new fishing vessels, which were under construction in 1986 and 1987 [167]. The reaction of the Community to the increased capacity during 1987 and 1988 was an *a posteriori* move to freeze all construction projects started in 1988 although the cause of the increase was the unfinished construction projects in 1986 and 1987.

Another serious defect for the implementation of the MAGPs was that they relied from the beginning on the information provided by Member States and not on an objective Community register of fishing vessels. Although MAGPs aimed to cover all the fishing capacity in the Community, there were Member States that overstated their capacity, for example the German data for the fishing fleet included boats for sport fishing. Others understated their fishing capacity including Greece whose data consisted of the inland fishing fleet only and Spain which excluded vessels in the "basic lists". Other Member States do not keep consistent statistical data for the overall capacity of their fishing fleet. For instance, data from Italy cover only licensed fishing vessels, data from the Netherlands cover the capacity of the fleet that catch species covered by the quota system. All Member States tend either not to keep consistent data or they fail to measure the engine capacity of fishing vessels. Although the Community seemed aware of this situation and adopted a regulation in 1986 [145] for all Member States to use identical rules to define the characteristics of fishing vessels, this rule was hardly implemented and the same situation continued until the end of the 1987-1991 MAGPs. In 1989 a regulation was adopted [150] that was to enforce standardization of the definitions and the characteristics of the fleet, whose application in practice will be discussed below.

The tonnage of the E.C. fishing fleet at the end of 1986 (first column, table 7.2) is believed to describe more realistically the capacity of the fishing fleet in GRT as it is based on revised data and has been released recently by the D.G.

Table 7.2   1986-1991 MAGPs, E.C. fishing fleet capacity in GRT units

	Situation at the end of 1986	E.C. position at 1.1.87 according to initial decision	E.C. position at 1.1.87 according to final decision	Target set for 31.12.90	Situation on 31.12.90	Target set for 31.12.91	Situation on 31.12.91
Belgium	22,846	22,846	25,165	22,870	25,498	21,340	27,089
Germany	59,000	49,900	51,500	50,120	45,837	76,950	48,910
Denmark	133,477	136,488	136,680	122,687	119,356	119,400	119,188
France	205,016	191,028	209,560	204,786	204,512	201,604	184,495
Greece	181,109	137,761	137,761	133,672	132,546	130,945	131,970
Ireland	41,000	53,065	58,845	49,903	51,017	43,941	50,693
Italy	284,631	262,717	302,986	282,114	277,563	268,198	268,198
Netherlands	126,632	77,500	82,400	71,840		64,796	82,400
U.K.	150,160	148,403	206,934	150,336	201,745	193,027	217,273
Spain	678,888	663,158	631,838	612,881	587,704	604,757	567,568
Portugal	194,971	210,525	208,670	209,540	166,032	209,540	182,829
Community	2,077,730	1,953,391	2,052,339	1,887,879	1,811,810	1,934,287	1,880,613

note: Data and targets from1986 to 1990 concern only F.D.R. under Germany; after 1990 they concern the unified Germany

sources: O.J. (31.12.1992) No L401 and Court of Auditors, Special report No 3/93 and D.G. XIV, XIV-A-3 30.6.1993

XIV. Comparing this to the data on which the initial decision for the setting up of the objectives of the 1987-1991 MAGPs was based, there is a considerable overall difference of 124,339 GRT which is, for example, the tonnage of the whole fleet of the Netherlands at that time.

Moreover, the Community's unconditional trust in the information provided by the Member States was carried through the years 1988 to 1991 and had as a result continuous revisions of the objectives of the 1987-1991 MAGPs. According to the final decision of 1988 [168] the ocean-going tuna fleet, fishing vessels used for transport, aquaculture and for fishing for molluscs and bivalves are excluded from the reduction in fleet target. Despite this exclusion the Community was faced with an increased capacity in the fishing fleet (see tables 7.2 and 7.3, third and second column respectively). Compared with what was recorded on the date of the initial decision there is a difference of 98,948 GRT (i.e. 5% more in tonnage capacity) and 870,391 KW (i.e. 11% more in engine capacity).

Although the interim objectives for 31.12.90 have been generally achieved, this success is spurious if the situation of the E.C. fleet at the end of 1991 is compared with that at the end of 1986 taking into consideration the unification of Germany. A comparison in terms of GRT, where there are realistic data for 1986, reveals an overall increase of 7.5% although the Community's objective was envisaged to be a decrease of at least 3% in tonnage capacity. The full objectives set for the end of 1991 in terms of tonnage have been more than achieved. However, in terms of engine power the target capacity has not been achieved, and the situation in the end of 1991 was 5.4% higher than the objective.

Additionally, the reliability of the 1991 data for the actual fishing fleet position remains questionable. For example, what is given by the Community as the situation at the end of 1991 did not take into consideration the complete lists of the fleet capacity of the former DDR. Indicative of the disposition of the Community to include hastily the former DDR in the MAGPs' objectives is the decision [153] taken two months before the end of the 1987-1991 MAGPs, which moved the target fleet capacity upwards by 27,750 GRT and 44,500 KW to cover the whole German fleet, while at the same time there is no definite estimate of the capacity of the former DDR fleet. Moreover, the situation of the Greek fleet at the end of 1991 is not at all clear. This is because the Greek data were inaccurate, excluding as they did the island fishery fleet. Furthermore, the existence of data showing the capacity of fishing fleets of Member States, which are estimates and extrapolations of previously recorded situations, present further complications. This applies literally to all Member States' fishing fleet engine capacity data.

Table 7.3 1986-1991 MAGPs, E.C. fishing fleet engine power capacity in KW units

	E.C. position at 1.1.87 according to initial decision	E.C. position at 1.1.87 according to final decision	Target set for 31.12.90	Situation on 31.12.90	Target set for 31.12.91	Situation on 31.12.91
Belgium	71,250	78,506	72,495	77,164	69,242	79,816
Germany	137,000	139,100	138,440	127,655	182,500	190,273
Denmark	571,996	563,667	524,515	506,929	514,716	488,278
France	918,087	1,158,576	1,095,460	1,140,196	1,055,050	1,189,849
Greece	568,823	568,823	523,795	576,297	493,776	710,899
Ireland	212,260	234,892	200,502	186,470	177,576	186,054
Italy	1,295,804	1,796,829	1,643,730	1,488,724	1,541,664	1,536,518
Netherlands	462,900	498,000	429,570	485,300	382,278	441,953
U.K.	759,953	1,155,212	1,119,206	1,173,548	1,095,206	1,228,922
Spain	2,130,156	1,831,554	1,778,667	1,721,148	1,756,001	1,671,551
Portugal	542,527	515,988	541,003	494,856	541,003	504,067
Community	7,670,756	8,541,147	8,067,383	7,978,287	7,809,012	8,228,180

note: Data and targets from 1986 to 1990 concern only F.D.R. under Germany; after 1990 they concern the unified Germany

sources: O.J. (31.12.1992) No L401, Court of Auditors, Special report No 3/93 and PE 140.600, 7.1991

Therefore, for the series of reasons given above, the uncertainty and the limited degree of reliability of the results of the MAGPs are what remains as a striking conclusion from the adoption of the programmes. Although the Community tried to improve the monitoring of fishing capacity by introducing a Community register system, this came too late (in 1989) and had no effect on the realisation of the objectives discussed above.

Critique of the Community fishing vessels' register

In spite of two regulations aimed at the definition and standardization of fishing vessels and which were supposed to lead to a Community register of fishing vessels in 1983 [145] and 1986 [146], the Community has remained unable to enforce such a system. The Community's inertia in implementing such a system in a concrete form on one hand and the reluctance of Member States to provide consistent and correct data on the other have led to the discontinuity and delay in the application of the envisaged Community register system. At length, by the regulation of 1989 the Commission requested the application of such a register system by 30 September 1989 and committed an amount of 490,000 ECUs for data processing services and advice. By the beginning of March 1990 all Member States had taken the measures needed for the application of the system. As a result of administrative measures that empower the transferability of information between the Commission and the Member State concerned and also authorise the investigation and the inspection through censuses and audits carried out by Commission officials, the Community register system for fishing vessels has been fully applied since the end of June 1991. Then, for the first time Member States forwarded to the Commission a magnetic data-carrier with information about their fishing fleet capacity. It is because of the application of this system that the unreliability of the information about the capacity of the fishing fleet, which the Community possessed during the 1983-1986 and 1987-1991 MAGPs, has been revealed, corrected and revised.

In spite of the establishment of the Community register of fishing vessels, there are still disparities in the measurement of the fishing capacity between Member States. Under the 1986 regulation [145] two parameters are used to measure the fishing capacity of a vessel, namely the tonnage and the engine power. Regarding the tonnage there are several systems of measurement that are applied in Member States. There are two main measurement systems, the "Gross Tonnage" unit as it had been specified in the 1969 London Convention on the tonnage measurement of ships and the system defined by the Oslo

Convention where the unit of measurement is defined as Gross Registered Tonne (GRT). Under the 1986 regulation the tonnage measurement system defined in the London Convention will be obligatory for all vessels after the 18 July 1994. Before that, the GRT measurement continues to be valid.

In practice, although Germany, Denmark and the Netherlands express the tonnage of their vessels according to calculations of the London convention others employ this method only for vessels over 24 metres long. For smaller ships these Member States use other customary criteria or calculate the tonnage capacity of their fleet using a simplified formula as compared to that defined by the Oslo convention for GRT. As the Court of Auditors notes [154] there are almost identical fishing vessels in the Community which although they were built in the same shipyard, because they are used by different Member States, they can have a difference of 60% in the tonnage register even though their tonnage capacity is measured in GRT.

As a result, the comparison of table figures totals remains only indicative of how the tonnage capacity has changed in the Community and conttributes little to the knowledge of the true and absolute fishing capacity. In addition, this situation could have a detrimental effect on a system of licences for fishing vessels that the Community envisages applying in the future, not later than 1.1.1995 [155]. This has to be settled before the application of a licence system before 2002 when the Community fishing grounds are scheduled to become freely accessible to all Member States. The establishment of a standardized system for measuring tonnage capacity is of vital importance and represents the corner stone for the application of the future MAGPs as well as for the co-ordination and management of all the structural aid measures of the Community. Instead of subsidising the modernization of vessels it is vital for the Community to modernize its knowledge about the fishing capacity, and if necessary by meeting part of the costs for re-measuring the capacity of fishing fleet in a standard way and by enforcing a uniformity of vessel measures throughout the Community. A discussion of the costs of the Community's structural measures for the fishing industry is given below.

Regarding the power criterion of fishing vessels, this is defined by the 1986 regulation [145] and is measured as the total of the maximum continuous power, determined according to the International Organization for Standardization (ISO) using the Kilowatt (KW) unit. The data on the power capacity of the fishing fleet for all Member States, given by sources such as the EUROSTAT, FAO and OECD over the years, are discontinuous erratic or roughly estimated and are presented with caution about their reliability.

The importance of the engine power criterion for measuring the fleet capacity in relation to the aim set by the Community for the structural policy is ambiguous. Certainly this criterion is significant for vessels employed in trawling but it is far less important for vessels which are employed in other fishing methods such as using lines and hooks and for fishing with creels and pots to catch crustaceans. For the latter, it is senseless to be obliged to follow the Community's policy of reducing engine capacity, since it would present a disadvantage with respect to safety in situations of emergency at sea.

A common practice for fishing vessels is to reduce engine capacity by fitting a derating device to the engine in order to comply with the measures for power limitation. However, how long this device remains fitted is disputable, since it can be easily removed. Moreover, it would be economically inefficient for the Community to grant aid for shipbuilding and modernizing vessels, in accordance to the engine power reduction policy, for such fishing vessels that are employed to the type of fishing where engine power is not important. This is the case of long lines, potting and creeling, drift-netting and purse-seining where vessel velocity is not directly related to the fishing capacity. Therefore, in addition to the establishment of consistent and complete information about the engine power capacity of the Community's fishing fleet there should be a more reasonable use of this criterion towards the objectives of the MAGPs. There is a requirement for a weighted engine capacity measurement for fishing vessels according to the fishing method that they employ.

The implementation of the Community's structural measures

The structural regulation of 1986 [145] as it was amended in 1990 [147] provides for Community aid for a series of structural measures that can be grouped into the following categories:

1 Measures to improve infrastructures i.e. financing for the construction and modernization of vessels and developing port facilities.
2 Measures to reduce fleet capacity in Community waters i.e. schemes for temporary or permanent withdrawal of vessels, re-deployment operations, exploratory fishing and joint ventures and other specific measures.

Construction and modernization of fishing vessels

Before granting aid for projects under the first category of measures there must be an investigation to establish if the project conforms with the objectives of the MAGPs according to the 1986 regulation [146] articles 6(2)a and 9(2)b.

However, it is not clear which are the absolute guidelines that define if the project is in accordance with the general objectives of tonnage and engine capacity of MAGPs and furthermore, which authority is responsible for monitoring that the project conforms with the MAGPs. This is because there is no regulatory provision for this in the E.C. structural regulations for the fishing industry. The procedure and the final decision for the approval or rejection of a project on the ground of conformity with the MAGPs objectives is reminiscent of a Kafka novel, since not only the rules but also the authority that applies the rules are obscure and unknown [154]. Given also the unreliability of the information that the Community acquires on the capacity of the fishing fleet and, furthermore, considering the undefined "intelligent guess" that some authorities have to make in order to find out if the project is in compliance with the MAGPs objectives before the end of the period of a MAGP, it is unlikely that the decision would be a justified one. In practice, increases in fishing capacity are permitted by Commission decisions in exchange for the withdrawal of capacity and, furthermore, the Commission ensures that before granting financial assistance "a link" is always established between the decision to invest and the compliance of Member States with the aims of the MAGPs. On this ground the Commission justifies [157] a refusal to finance projects proposed by the U.K., submitted in 1992, since this Member State had not met the objectives of the 1987-1991 MAGPs.

It is the Commission's responsibility to monitor the implementation of the MAGPs and in addition it is the Commission's responsibility to calculate first, before any action is taken by the Member State concerned, the amount of aid for ship-building and for modernization projects (articles 7(1) and 10(1) respectively of the 1986 regulation [146]). This procedure for the calculation is made under rules found in Annex II of the above regulation, and this calculation is supposed to take place after the project has been found to comply with the objectives of the MAGP. Then the amount for the Member State's contribution is fixed accordingly. Nevertheless, in practice this is not always the rule. For all the construction and modernization projects for fishing vessels in Germany, France, Ireland and the U.K. in the years 1987-1990, national aid had been granted before the Commission had made its decision. Moreover, the national aid granted in Ireland and in the U.K. was without taking into consideration the Community's target for the 1987-1991 MAGPs since both Member States had not achieved these objectives at the end of 1991 (see tables 7.2 and 7.3 pages 206 and 208).

The flow of Community expenditure for shipbuilding and the modernization of fishing vessels is shown in figure 7.1. Unfortunately, there is no separation of the Community's expenditure between the construction and the

213

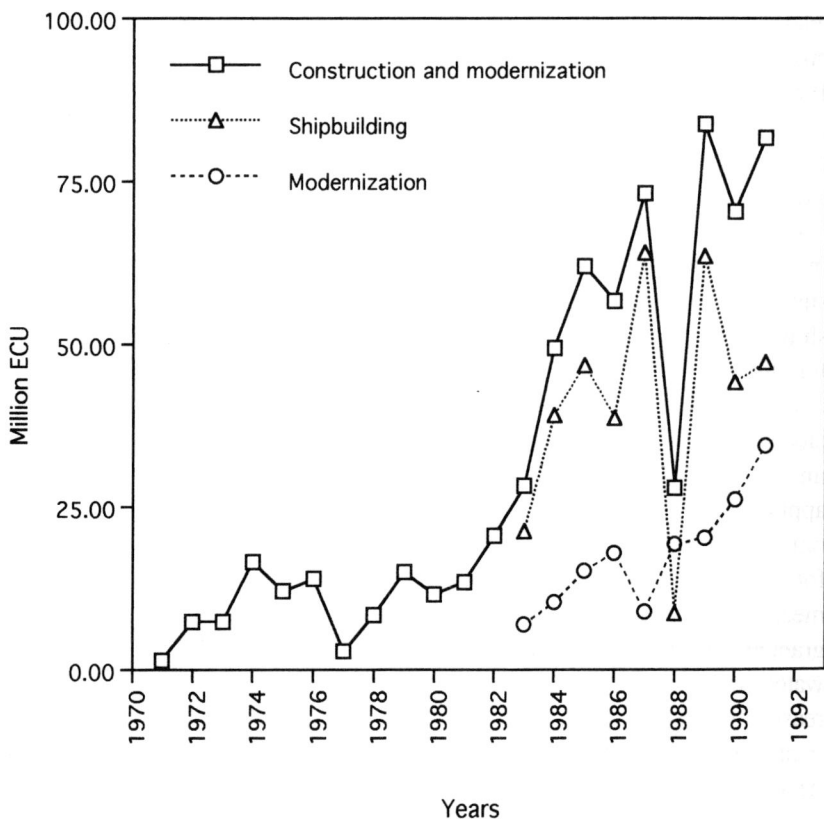

note: Data 1971-1972 for 6 Member States, data 1973-1980 for 9 Member States, data 1981-1985 for 10 Member States, data 1986-1992 for 12 Member States

Figure 7.1   E.C. annual expenditure on the construction and modernization of the fishing fleet

sources: D.G. XIV, XIV-A-3 30.6.1993, SEC(91) 2288 (final) 18.12.1991 and O.J. (15.12.1992) C330/107

modernization of fishing vessels before 1983. For the years of the first MAGP (1983-1986), although there is such separation (Figure 7.1) there are no complete data of the number and the kinds of projects that received aid from the Community. Although for the second MAGP (1987-1991) the Community has released data for the projects, these have to be examined sceptically, since these projects have not been confirmed by the Member States concerned. Under the 1986 structural regulation Member States have to produce a report stating the projects which received aid and the results of each of them. These reports have to be produced at most two years after the financial aid is received. With the exception of the U.K. no Member State has produced such a report.

Although the Commission has to decide twice a year whether to grant aid for the construction and modernization of vessels, this decision is usually significantly delayed. For example, the sharp drop in aid given in 1988 for shipbuilding (Figure 7.1) is likely to be attributed to a change in the guidelines for shipbuilding adopted by the Community [158], which exclude the granting of aid for fishing boats that are intended to fish in Community waters; also to a change in the submission of applications and administrative rules for projects under the regulation 116/88 of 20 April 1988. Although in 1988 there were applications for projects from all the Member States, the Commission accepted only projects from Portugal [167]. Also, probably in order to give a justification for a decision [156] approved at that time, namely to apply the structural measures of the E.C. to the French overseas departments, the Commission granted aid for shipbuilding for tuna vessels intended to fish in tropical Atlantic waters and in the Indian Ocean. In the meantime the Commission states that the reason for the delays lies in "following internal rules ... that ensure equal treatment for all applications" [157]. In addition, the Commission asserts that decisions are delayed due to the lack of information concerning the state of progress of the MAGPs, given the need to ensure that the projects are in compliance with the objectives of the MAGPs. There is certainly insufficient and imprecise information concerning the capacity of the E.C. fishing fleet as analysed above. However, this should have been realised by the Community long before the application of MAGPs and unquestionably long before accepting proposals for the construction and modernization projects for fishing vessels. Thus *de facto* the Community has to act and decide on projects in a situation where information is inadequate.

The Court of Auditors has found cases where Community aid was granted after the project was completed. In addition there is a series of shipbuilding projects where aid was granted without careful consideration by the Commission [154]. Such cases are presented in a report by the Court of Auditors and they refer to cases where, for example, the Community granted aid without the beneficiaries having enclosed any estimate of the cost of the proposed project as required (particularly for projects from Italy and Spain, for example, project ES 190/87/1). They also point to project proposals which enclosed imprecise and unclear information about the tonnage and engine capacity of the vessel (projects IT 47/87/1, UK 253/89/2). In other cases aid was granted irrespective of what the request was about in the first place (projects F/1987, P 54/87/1) and for strictly identical project estimates the aid granted has differed (ES 166 and 173/87/1, IT 99 and 100/89/2). In addition, the Court of Auditors reveals numerous cases of vessel construction where vessels meant to engage in a particular type of fishing have obtained Community aid and were then replaced by other shipbuilding projects for vessels intended for a different type of fishing, although this is forbidden (for example, projects F 238/89/1 and PO 18 and 19 87/1). Others obtained aid without being eligible because increases in capacity were endorsed by Community decisions (for example projects (UK 151/87/1, I 51/87/1).

During the 1987-1991 MAGPs the Community selected 672 projects for shipbuilding which represent a tonnage capacity of 69,800 GRT and 238,000 KW in engine power (Table 7.4). The shipbuilding projects represent about 3% of the total capacity in the Community in terms of GRT and KW and were granted by the Community total financial aid of 177.3 million ECU. The Netherlands and Denmark do not have any national scheme for financial aid for shipbuilding and, therefore, they are not eligible for Community aid. Since 54% of the shipbuilding projects (365 projects) were granted aid during 1989 to 1990 only 307 out of a total of 672 were completed by 31.12.1991, that is 45.5%, 38 have been abandoned that is 6% and the rest are to be completed, that is 49% of the total number of projects. The utilization of the Community aid for all Member States exceeds 90% of the aid granted and the utilization of the total aid by completed projects, in terms of payments already made, is 54%.

Regarding Community aid for the modernization of fishing vessels this is granted only if the project is not contributing to the capacity defined in MAGPs. In addition to the argument that there are no defined limits on how a project can or cannot be in compliance with the MAGPs, a problem which in practice is left to the intelligent guess of the Commission which finally has to approve or reject the project, there is also the following vacuum in the E.C. regulation on this matter. Given the aim of the structural policy to adjust capacity to the fishing stocks, the Community presumes that only the tonnage and engine capacity of fishing vessels define the effort devoted to fishing and excludes projects, which would increase these two types of fishing capacity, while at the same time permitting projects that seek finance for electronic equipment and fishing gear in accordance with to the 1990 structural regulation [147]. Furthermore, under

Table 7.4 Shipbuilding projects funded in the period of the 1987-1991 MAGPs

	Commitments				Completed		Abandoned		To be completed	
	N	1000 ECU	GRT	KW	N	1000 ECU	N	1000 ECU	N	1000 ECU
Belgium	2	639.7	379	1,100	2	627.7	0	0	0	0
Germany	31	8,926.5	7,195	13,633	14	6,155.4	3	241.2	14	2,486.5
Greece	49	6,517.5	2,507	10,966	32	3,840.4	4	751.8	13	1,763.1
Spain	192	60,853.3	21,727	67,344	92	35,842.7	8	2,294.1	92	22,426.5
France	122	32,617.3	11,347	45,316	68	20,982.1	6	923.9	48	10,425.8
Ireland	15	2,052.9	420	2,484	6	572.2	0	0	9	1,459.8
Italy	119	27,257.2	8,523	41,871	31	9,323.8	1	22.6	87	17,736.6
Portugal	116	32,055.9	15,158	44,525	40	12,604.4	11	2,962	63	15,929.6
U.K.	26	6,369.6	2,504	10,761	22	5,255.7	3	196.4	1	801.3
Total	672	177,289.7	69,800	238,000	307	95,204.4	38	7,391.9	327	73,029.2

sources: Court of Auditors, Special report No 3/93 and PE 140.600, 7.1991

the 1990 structural regulation on aid for the modernization of vessels, small vessels between 5 and 12 metres long are also included in aid provisions. In this way the Community allows the installation of more effective fishing equipment and electronic equipment for the detection and location of fish stocks, both of which contribute to an increase in the fishing efficiency of a vessel and therefore to the fishing capacity of the fishing fleet.

During the period 1987-1991 of the MAGPs the Commission was committed to grant 75.5 million ECU of aid to 2,214 projects for the modernization of fishing vessels (Table 7.5). 1,372 (62%) of the projects were completed by the end of 1991 and utilized 38.4 million ECUs, that is 51% of the Community aid. 2.7 million ECUs were unused, although they were allocated by the Community to these projects. This means that on average each project was completed by utilizing 2,000 ECUs less than the average allocation of aid by the Community. The projects that were abandoned before they were finished represent 6.1% of the total projects and had been allocated 3.4 million ECUs, that is 4.4% of the aid committed. Given the small percentage of abandoned projects, the Member States have used more than 93% of the Community aid. The Netherlands and Denmark have the highest rate of abandoned projects in the Community (15% and 25% of abandoned projects respectively). This may be due to strong permanent programmes for the withdrawal of capacity, which exist in these two Member States and result in the highest rates of withdrawals in the Community, as will be discussed further below.

Projects eligible for Community aid in the structural policy must also comply with the provisions of the national aid system of Member States according to the 1986 [146] structural regulation as amended in the 1990 [147]. In the case of the modernization of fishing vessels only Netherlands and Denmark exclude from their national aid scheme modernization projects which contribute to the increase in the fishing effort such as electronic devices for detection and fishing gear. Although this decision is in accordance with the Community's aim to reduce fishing capacity it introduces unfairness into of the Community's treatment of the modernization projects of these two Member States compared to other Member States that provide national aid for detection equipment and fishing gear.

A fairer treatment of this problem from the Community's point of view, which will also be in accordance with the objectives of the structural regulation, would be as follows. First, the Community should re-define fishing capacity to include more elements than the engine power and tonnage capacity found in the MAGPs, and these should be weighed according to their contribution to the fishing effort. For example, these elements need to include detection devices, fishing gear, type of fishing method and the region that a vessel is routinely

Table 7.5 Modernization projects funded in the period of the 1987-1991 MAGPs

	Commitments		Completed		Released	Abandoned		To be completed	
	N	1000 ECU	N	1000 ECU	1000 ECU	N	1000 ECU	N	1000 ECU
Belgium	35	951.2	29	677.1	10.2	1	6.2	5	257.7
Denmark	239	5,871.3	203	4,632.1	270.5	36	968.7	0	0
Germany	79	1,294.3	46	779.6	154.3	2	29	31	331.6
Greece	102	2,131	77	1,477.7	76.5	9	99.8	16	477
Spain	714	26,344.5	482	15,136.4	1,090	36	1,224.1	196	8,894.1
France	110	4,344.5	48	1,996	70.1	1	73.7	61	2,204.8
Ireland	119	3,694.3	96	2,386.3	208.5	8	86.7	15	1,012.8
Italy	385	19,547.1	74	3,068.6	555.9	5	204.4	307	15,718.2
Netherlands	48	621.6	21	283.2	19.6	12	179.1	15	139.7
Portugal	112	3,914.4	57	1,899	103	10	252.3	45	1,660.8
U.K.	270	6,778.8	239	6,090.1	152.9	16	230.4	15	305.2
Total	2,214	75,493.1	1,372	38,426.2	2,711.4	136	3,354.3	706	31,001.8

sources: Court of Auditors, Special report No 3/93 and P.E. 140.600, 7.1991

fishing in relation to the availability of fishing stocks in the region. Secondly, the Community should not exclude all the modernization projects of a Member State that fail to meet the objectives of a MAGP. There are investments that modernize a fishing vessel without increasing its capacity. For example, measures for safety on the board and measures that contribute to the better quality of fish landed such as storage, processing and equipment for packing on board.

Measures to reduce fleet capacity in community waters

The measures which the Community enforces to adjust fishing capacity to the catch potential concern schemes for the temporary or permanent withdrawal of vessels, re-deployment operations, exploratory fishing and joint ventures and other specific measures [146]. Although the Council in 1983 adopted a directive [144] on measures to adjust fishing capacity to catch potential, these are not likely to have received a proper consideration since there are no data available on the breakdown of the Community's expenditure for the various structural measures and the overall expenditure appears to have remained exactly the same for the years 1983 to 1986 at 6.86 million ECUs per year [58]. This supports the argument that decisions were taken on a case by case basis with no overall strategy and global targets behind them. The 1986 structural regulation (146) was a more decisive move in the application of these measures.

Permanent and temporary withdrawal of fishing vessels

The 1986 structural regulation [146], as it has been amended by the 1990 regulation [147], stipulates that in the event of a temporary decrease in fish stocks, a temporary withdrawal premium for fishing vessels applies until the stocks are replenished. Furthermore, if the reduction of stocks is to be persistent a final cessation premium is granted for vessels over 12 metres long which are engaged for at least 100 days per year in fishing activity. Permanent withdrawal of a fishing vessel means that either the vessel must be dismantled or transferred to a third country or used for non-fishing activity in the Community's waters. In addition, the permanent withdrawal of vessels according to article 8(2) of the 1986 regulation was a priority criterion for a shipbuilding project. Given the objectives of the MAGPs for capacity reduction this criterion becomes the principal one for the selection and the final decision on shipbuilding projects. In addition, because of technological progress the Commission [157] requires that withdrawn capacity is not replaced with an equal potential so that when selecting projects for shipbuilding the Commission has established a priority order insisting that the rate of withdrawn capacity be more than 100% of to the new shipbuilding project.

However, a different Community performance in decision making is seen when comparing the figures in table 7.6 for the period of the MAGPs 1987 to 1991. Overall, although the tonnage withdrawal is 3% more than the shipbuilding tonnage capacity, in terms of engine capacity there is an 11% increase in capacity for shipbuilding projects. The rate of withdrawal varies significantly between Member States, so that the overall estimation is a crude indicator. Portugal and Spain have the highest withdrawal rates of 166% and 114% in tonnage and 126% and 109% in excess of the capacity of shipbuilding projects, respectively. All the other Member States have built more vessels than they withdrew in terms of tonnage and engine capacity, with Ireland having the lowest rate of withdrawn capacity, 9% in tonnage and 12% in engine capacity.

In the absence of data that show the effectiveness of the newly built fishing capacity compared to the old fleet, it is not possible to determine exactly what must be the rate of withdrawals in exchange for newly built fishing capacity. There is a general belief in Community reports that technological progress in the fisheries sector accounts for about 2% of the annual increase in fishing effort. Whether such a measurement can be reliable, cannot be asserted but it is undeniably true that a new vessel with exactly the same tonnage and engine capacity as an older one can be used to fish more efficiently than the older one, as technology and vessel design are progressing fast. Given also that out of 725 fishing vessels withdrawn in the Community during the 1987-1991 MAGPs, 616 (that is 85%) were built more than 20 years ago, this contributes to the argument that the withdrawal rate must be more than one to one for a reduction in the Community's fishing capacity.

Moreover, careful consideration must be given in assessing the fishing possibilities of a new vessel as to which area and on which stocks the new one will exert its effort compared to the withdrawn vessel. For example, there should be a balance between the fishing rights and the kinds of licence granted to a new vessel and those enjoyed by the withdrawn vessel. Furthermore, although the Community has allocated about one fourth of the expenditure for structural policy since 1983 to "adjustment of capacity" measures, there is no provision in the CFP on how fishing activity is to be defined, which by definition must be the measure and the degree of the usage of fishing capacity. Above the physical characteristics of a fishing vessel (GRT and engine power) and the fishing rights or licences that a vessel may have, what determines the pressure of this capacity on the stocks is the fishing activity of the vessel. Therefore, the Community before attempting to adjust capacity must define a

Table 7.6   E.C. Fishing fleet capacity constructed and withdrawn during 1987-1991

	Shipbuilding GRT	Permanent withdrawal GRT	Shipbuilding KW	Permanent withdrawal KW
Belgium	379	247	1,100	941
Germany	7,195	2,407	13,633	6,749
Greece	2,507	1,683	10,966	7,237
Spain	21,727	24,885	67,344	73,165
France	11,347	10,404	45,316	34,861
Ireland	420	41	2,484	308
Italy	8,523	5,973	41,871	30,332
Portugal	15,158	25,117	44,525	56,415
U.K.	2,504	1,139	10,761	3,976
Total	69,800	71,896	238,000	213,984

note: The Netherlands and Denmark are not included in the list because their national legislation does not provide aid for shipbuilding

source: COM SEC (91) 2288 and Court of Auditors, Special report No 3/93

relationship between the capacity and the pressure that this exerts on the fish stocks. The key parameter in such a relationship is fishing activity.

What happens in practice tends to be more perplexing. The Court of Auditors presents a series of accusations that the Commission does not take proper consideration in the assessment and confirmation of files for shipbuilding projects and concomitant withdrawals [154]. For example, these accusations start from the point that aid has been granted for projects that do not specify the physical characteristics of the new vessel, that is GRT and KW and ends at the other extreme that Commission has accepted as concomitant withdrawals, vessels which had already sunk long before the 1986 regulation [146] that allows for such transactions (projects P 11/88/2, GR 19/89/1). Moreover, withdrawals have been accepted even though there was no evidence that the vessels concerned were fishing vessels (projects ES 88/87/1, P 10/88/2, IT 262/89/2).

The Community's expenditure on the permanent withdrawal of fishing vessels was 97.9 million ECUs in commitment appropriations for the period 1987 to 1991. Overall, as table 7.7 shows, out of this amount 52 million ECUs (that is 53%) were used to withdraw capacity in the E.C.. Ireland and the U.K. failed to avail themselves of such Community aid. Belgium and France used only 5% and 13% respectively of the aid available for them, while of the other Member States Germany, Portugal and Denmark utilised the highest percentages of the aid available to them with 79%, 78% and 72% respectively. In terms of withdrawn tonnage capacity Denmark, the Netherlands, Spain and Portugal withdrew a greater tonnage capacity than the other Member States.

Regarding the temporary withdrawal premium, this is granted to certain vessels (at least 18 metres long) which suspend their fishing activities for at least 45 days per year more than the average number of days of inactivity determined during the three previous years. In addition to the fact that there is not a unified definition of what fishing activity means, there is another question about the relation of this measure to its objective. That is, according to the 1986 regulation [146] this aid is to be granted until a temporary fall in the fish stock is reversed. How this objective relates to the 45 days cessation of an undefined fishing activity of a vessel remains a mystery and is left to the Member States to apply.

The fishing activity of a vessel can be separated into two distinct categories. First, the type of fishing effort, which is a combination of parameters that directly cause fishing mortality, for example the fishing method, the time of dragging the trawl, the engine power, the effectiveness of electronic locating devices. Second, the type of the fishing effort which is the concern of economics and includes all the other activities that are important for the

Table 7.7   E.C. expenditure for permanent capacity
             withdrawal and tonnage withdrawn for
             the fishing fleet during 1987-1991

	Commitments (1000 ECU)	Paid (1000 ECU)	Unused (1000 ECU)	Withdrawn GRT
Belgium	1080	57	1,023	165
Denmark	21,919	15,865	6,054	16,081
Germany	1,952	1,550	281	1,825
Greece	9,173	3,218	5,955	6,735
Spain	15,450	7,778	7,672	15,281
France	1,094	152	942	151
Italy	14,286	6,852	7,434	8,740
Netherlands	24,973	10,300	14,673	15,980
Portugal	7,956	6,239	1,717	13,686
Ireland	0	0	0	420
U.K.	0	0	0	1,139
Total	97,883	52,011	45,872	80,203 [2]

source: Court of Auditors, Special report No 3/93 and PE 140.600, 7.1991

economic success of any fishing operation. That is all these operations that produce additional costs or increase revenue. For example, indicators would be the time spent searching for fish, the time spent for transporting the fish between fishing ground and port, the handling facilities on board for the catch. Fishing vessels are likely to suspend their activities and remain in the port at a certain ratio of the above defined fishing activities. In addition, other sociological parameters, such as religion and ethics (e.g. Sundays, holidays), social status in a community, the family business status, alternative employment etc., can affect fishing activity beyond what would be expected in a rational economic relationship. Therefore, fishing activity is determined by the balance of many parameters, only some of which are more or less related to biological criteria. Consequently, by providing an unrefined measurement, such as the time when fishing activities are suspended (with the aim of replenishing fish stocks), this takes no account of the validity of all the factors affecting the situation.

In the absence of such crucial definitions and appropriate measures which would objectively lead to a resolution, the aid that is given for the temporary withdrawal of fishing vessels becomes mainly a welfare measure for fishermen which burdens the European taxpayer. However, only 41% of the eligible aid has been paid out between 1987 and 1991. Since the application of this measure in 1986 the Community has each year spent more than 21% of the total expenditure on withdrawals, a total during the 1987-1991 MAGPs of 167 million ECUs. Spain has received the most of this type of aid with more than 80% of the total paid amount, that is 14.2 million ECUs [169].

Overall assessment, future perspective

The whole structure of the Common Structural Policy for the fishing industry has been based upon the overused words, "overcapacity in the fishing industry". This mysterious concept of overcapacity is not only the sugar plum in every Community text about the fishing industry but also the scapegoat in all articles about the E.C. fishing industry, and one which needs to be exorcised. In attempting to do this there is an overwhelming effort to measure it first, and then declare a percentage which must be withdrawn depending on the objectives of the writer. According to a report produced by an independent committee under the aegis of the Community [159] it has been estimated that at least an average 40% fishing fleet capacity reduction should be achieved in order to reach the biological objective of the MSY and possibly more must be reduced to achieve the MEY. Others in the Community envisage that an overall

90% reduction should be achieved [160]. Fishermen seem to accept the 40% proposed reduction without any scepticism [161].

It seems that there are more than the biological uncertainties (explained in the second chapter) of attempting to relate a substantial reduction of fishing effort to a single fish stock population and the unreliable, or absent, data for catch per unit of fishing effort that should be taken into consideration. In practice the problem concerns a more perplexing situation of multi-species and multi-fleet fisheries. This is because a gear can catch various species and a vessel can operate various gears. Kirkley and Strand (1988) reject the assumptions of the independence in inputs and the separability of outputs made by fisheries managers. This means that there are technical and economic interactions in production and single stock or species management would affect the exploitation of other species in a multi-species fishery. Another complication that is revealed in their work is that regulations imposed independently of the technology of each tonnage class of vessel or of recognition of different operational strategies are likely to have different impacts on fishing firms and the fish stocks. This work represents progress over traditional thinking and demonstrates that different types of management and regulations may be necessary if a fleet is comprised of heterogeneous fishing firms. Such a view can contribute to effective decision-making in fisheries management.

In addition, there is a series of economic and socioeconomic problems which further complicate the situation. For instance, in attempting to measure the effects of fishing overcapacity in the economy one has to take into consideration not only the increased costs and the reduced income for the fishing industry but also the effect on society as a whole. The task of devising solutions is even more complicated. Fishery economists have given a number of solutions (e.g. limitation of fishing input, regulation of effort through licences, limitation of access) which typically rely on the maximization of the resource rent in fisheries. In considering the complicated nature of the problem, any absolute declarations for fishing capacity reduction have to be taken as indicative measures since they are usually made under certain assumptions and aim to fulfil different objectives. For example, how can individual fishermen be forced out of business when the fleet is considered exceedingly large. The measures that the Community envisage to obtain such a reduction have ambiguous results. On the other hand a free market solution will have undesirable results since in a stock depletion situation even small catches at high prices can sustain a substantial part of the fleet in an economically viable position. Moreover, even if a drastic reduction of the fleet occurs, what will the consequences be if the remaining vessels fish more intensively and exert more pressure on the stocks. In addition, how will other sectors of the economy be affected by the unemployment which arises in the fishing industry. It can be argued that a reduction in vessels with the same catches from the same fish populations would reduce the average cost of fish, thereby helping the processing industry and possibly increasing the quantity of fish demanded at a lower price. While this scenario is feasible it will still lead to a reduction in employment at sea and shipbuilding and ship repairing.

In practice, the Community's structural policy for the fishing industry carries a series of inadequacies. First of all, although the Community's principle objective is to reduce the fishing capacity of the fleet, the Community has ended up by granting more financial aid for the construction and modernization of vessels than for the withdrawal over the years 1971 to 1990 (see graph in figure 7.2). Also by comparing the average cost to the Community per GRT for withdrawal and for shipbuilding the Community's expenditure for a withdrawn GRT (Figure 7.2) is four times less than for building a GRT. This is to be expected, since the withdrawn vessel is less valuable than a new one with the same GRT units. However, it is less reasonable to follow such a policy when the objective is to reduce capacity. Moreover, such a policy becomes more inconsistent when the particular deficiencies of each measure, discussed above, are taken into consideration.

Because there is incomplete information on which to base a rationally planned reduction of the fishing fleet and because the Community decommissioning schemes are based on uncertainties with no prior robust analysis of outcomes, the only thing that remains is to learn and update our knowledge from their outcomes. To accomplish this it is of vital importance to establish a well planned monitoring system to enable us to see the impact of each measure. Under a monitoring system, in addition to collecting data on fish population, capacity of fishing vessels and their activity (fishing methods, time spent on fishing, landings etc.) one must also gather data on economic parameters, such as the breakdown of the costs of fishing effort. This will enable us to determine the relevant importance of the sectors of the fishing industry in a regional, national or international context.

However, such learning derived from trial and error does not automatically justify the extension of the measures that the Community has planned up to now. A typical example is the so-called "small scale pilot actions" for the re-deployment of fishermen. Although the Commission admits that there is no information that can be used to define the nature and even more the scope of such a measure [164] the same body grants half of the assistance, that is 50,000 ECUs, to be used for training fishermen for alternative employment. One such pilot action was undertaken in 1990 in Grampian region, in the port of Macduff which has a high dependency on fishing with around 600 people out of a

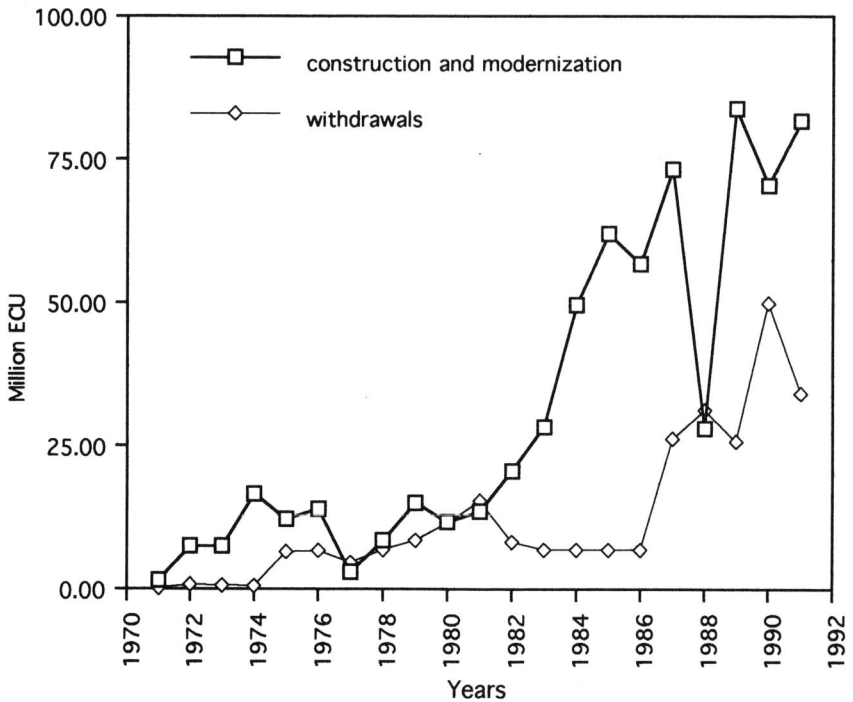

note: Data 1971-1972 for 6 Member States, data 1973-1980 for 9 Member States, data1981-1985 for 10 Member States, data 1986-1992 for 12 Member States

	Belg.	Den.	Ger.	Gre.	Sp.	Fr.	It.	Neth.	Port.	U.K.	Ir.	Average E.C.
W	346	987	849	478	509	1005	784	645	456	N/A	N/A	661
B	1,688	N/A	1,241	2,600	2,801	2,869	3,183	N/A	2,115	4,888	2,544	2,434

W: Average cost per withdrawn GRT for the Community during 1987-1991
B: Average cost per built GRT for the Community during 1987-1991

Figure 7.2   Expenditure on construction and modernization of fishing vessels and on permanent and temporary withdrawal measures

source: D.G. XIV, XIV-A-3 30.6.1993, SEC(91) 2288 (final) 18.12.1991, O.J. (15.12.1992) C330/107 and Court of Auditors, Special report No 3/93

228

population of 4,050 employed in the fishing industry (35% of total employment). Out of the 600 people only 250 were fishermen at that time [165]. The project ran for 12 months and commenced on 1 September 1991. During that time nine fishermen found alternative employment and benefited from the financial aid of this project. Another two ex-fishermen were retrained to work in the fish processing industry. The project failed to convince people who worked in the fishing industry to take early retirement. Although in the final report on this project [166] the responsible bodies felt proud of the results there still remains a series of unanswered questions. For example, there was a decline of 30% in fishing vessels registered in Macduff from 1980 to 1990 and consequently the number of fishermen must have declined. However, there is no analysis of whether the project really contributed to the decision of fishermen to find alternative employment or if, as one of them states, in an eulogising leaflet about the project, "the project came at exactly the right time" for him which maybe means that he was anyway about to leave and the aid came at the right time. Furthermore, there is no assessment of the economic efficiency of the project since the whole aid was aimed to be distributed in one sector of the fishing industry in Macduff, and this sector employs only 41% of the total workforce in the industry. In addition, it would be interesting to know in a region that benefits from preferential treatment from the structural measures of the Community, what the impact of the project was on other sectors of the regional economy.

It is not the scope of this work to analyse such a project, which represented only 0.04% of the total expenditure of the Community for fisheries in that year (1990). It has been mentioned as an example which attempts to show that if the Community had from the beginning defined the scope of the project and if it was well planned to reveal analytically what were the results, it would provide invaluable information. Regardless of the success or not of the project an objective analysis of the results would have been indispensable for our better understanding of the complex interrelations between the sectors of the fishing industry and their relation to a local economy which is relying on fisheries.

An advantageous measure that the Community has to enforce is the standardization of fishing capacity measurements and the definition of fishing capacity and fishing effort. This has to be measured by a weighed figure which would combine not only the physical characteristics of vessels but also the type of fishing they are employed to do and the area that they usually fish. Such a measure, which could work under the framework of the Community's register system that is now in operation, would produce a uniform system for measuring fishing capacity throughout the Community and, therefore, would be the corner stone for more objective knowledge of the pressure that the fishing fleet applies to stocks. It could also be the base of a proposed fisheries policy that would be founded on a licence system, as will be discussed in the final chapter.

Ironically, in revising for the third time the 1986 structural regulation at the end of 1992 [161] the Community literally substituted for the word capacity, which appears in the 1986 regulation, the term fishing effort, which appears in the new regulation, avoiding once more a clear definition of the term. Thus, the new regulation has changed the phrase "adjustment of the capacity to the catch potential" to the "adjustment of fishing effort to be in line with the balanced exploitation of the fishery resources". At first glance this seems more thoughtful than the previous term. However, there is no provision that defines the balanced exploitation of the resource. The only new component of this regulation, i.e. "fishing effort", has nothing to do with the notion of fishing effort as it is defined above. In spite of the new regulation which defines effort as the product of capacity and activity per section of the fleet, capacity is that previously defined for a vessel as tonnage and engine capacity. Moreover, the measure of activity remains confusingly undefined. As far as the sections of the fleet are concerned, these can be assumed to be as they appear in the new MAGPs for 1992-1996 [162], that is 1) vessels for bottom trawling fishing for demersal stocks 2) dredgers and beam trawlers for benthic stocks and 3) all the other fishing vessels. The 1992-1996 MAGPs demand a 20% reduction in fishing capacity (i.e. tonnage and engine) for the first category of the fleet, 15% for the second and no reduction for the third category.

8 Reforming the CFP

Review on the main features of the CFP

This work has so far expanded on the particular elements which compose the Common Fisheries Policy. They have been described and analysed in detail and their main weaknesses exposed. This section reviews the main issues which can be summarised as follows.

Resource management and conservation policy

1   The TAC and quota system fails to achieve its principal aim, which is to maintain fish stocks and conserve the resource. In several species landings are continually decreasing, while, in addition, the system generates a series of common practices that are incompatible with the conservation objective. For example, overcapacity, paper quotas, excessive discarding, "black" fish market etc.

2   The TAC and quota system has not had the desired influence on the direction of economic development in the fishing industry. The system perpetuates a situation where private costs of fishing are lower than social costs and which results in excess fishing capacity and inefficiency.

3   The reasons for the inadequacies can be traced partly on our insufficient knowledge of the biological parameters and ecosystem characteristics that would enable an assessment of the effect of fishing on the fishing stocks. TACs and other technical measures have often proved inadequate since they fail to consider the interrelations between species in the food chain, as happens for example in the case of industrial species.

4   On the other hand, for the time being fisheries management has little to gain from the use of MATACs and/or MSTACs. If the Community is

determined to apply such measures it must also face the cost of data collection to meet the requirement of multi-parametric models. In addition there is a series of economic considerations on the effect of MATACs and MSTACs on market forces and fishermen's behaviour which will create undesirable side effects for resource conservation and socio-economic upheaval in an already suffering industry.

5   Political pressures usually lead to considerable alterations of the scientist's recommendations, resulting in regular overestimations of the resource capacity, as happens in the agreed TACs. Moreover, unforeseen situations such as quota hopping and flags of convenience are incompatible with the principle of relative stability and furthermore jeopardise the Member States' obligations under the MAGPs. Therefore, as the E.C. moves towards a single market, the national quota system not only loses its effectiveness as a conservation measure but also becomes incompatible with the Community Law.

The common organisation of the market

1   The Community intervention scheme, which aims to ensure fishermen's income and the availability of supplies at reasonable prices, is unlikely to contribute significantly to the price formation or the volatility of prices of fish. Moreover, the system has failed completely to offer any protection to fishermen when crises from cheap imports have arisen. The inefficiency of the pricing mechanisms for producers' protection in fishing is seen in the inflexibility of the system which cannot track import prices fast enough especially when gluts of low-priced imported fish enter E.C. below reference price levels. Its cumbersome procedures which use up precious time before a Council decision can be made.

2   The policy of encouraging the formation of producers' organizations is inconsistent with that of maximizing social welfare from fisheries. It is unclear why the Community encourages the formation of Producers' Organizations since the objective of ensuring the income of fishermen has not been attained. It is unlikely that in devising fisheries policy any attention was paid to the principles of fisheries economics.

3   The Common Marketing Standards result in a distortion of the market, while keeping in existence a confusing nomenclature of standards for fishery products and generating a cumbersome administrative system which causes unnecessary complications.

4   Given that the Community is not self-sufficient and relies greatly on imports of fish products the effect of trade policy is questionable. Since

trade shows a yearly increase of 15% to 20% in the structural deficit since 1983 a removal of all trade barriers could benefit Community consumers and the processing industry from cheaper imports while it would provide an incentive to producers to be more competitive.

5   The benefits for the consumer and the processing industry that arise from the liberalisation of the trading regime with third countries established under GATT are dissipated. Consumers are obliged by the trade policy to be indirectly taxed in order to support this system for the benefit of the nation that exports the fish. Furthermore, the net result of the system of subsidies under the trade policy is to underline the inefficiency in the Community's fishing industry in comparison to others which can afford to sell cheaper fish products.

6   The high Community expenditure on third Country agreements is not matched by the benefits for the Community's fishing industry. Given the state of the resource in Community waters, the dismantling of customs duties and the Community's reliance on imports, more effort has to be exerted as a means of bargaining for fishing rights in third countries' waters in return for access to the Community market. The commercial character of such agreements must be the salient feature for economic efficiency as opposed to their current partial financial aid assistance character for the mixed type agreements and these with the ACP countries.

The Common Structural Policy

1   MAGPs aimed to adapt in a comprehensive way the Community fleet to the available resources. Unfortunately, it does not seem that the main objectives have been fulfilled up to now. Moreover, aid for modernization of the fleet and for construction of new vessels contributes to the expansion of the existing fishing capacity and adds to the current inefficiency of the fishing industry.

2   The main contributing factor can be traced in the inaccuracy of the information about the fishing capacity in all Member States which was the basis for establishing objectives for reducing fishing capacity. The main reason for this is that the Community relied from the beginning on unchecked information from Member States and not on an objective Community register of fishing vessels.

3   Although the Community recognised this inefficiency and aimed to contribute to the fulfilment of the MAGPs' objectives by a series of decisions and amendments, their effectiveness is dubious since they were

adopted after the setting up of the objectives and tried hastily to amend previously unforeseen situations. For a series of reasons the uncertainty and the limited degree of reliability of the results of the MAGPs are what remains as a striking conclusion from the adoption of the programmes.

4   An attempt to improve the monitoring of fishing capacity by introducing a Community register system came too late (in 1989) and had no effect on the realisation of the MAGPs objectives. In addition, this system did not enforce rigid and identical rules for defining the characteristics of the fishing vessels. This situation becomes subsequently burdened by a lack of a more comprehensive identification of the parameters that should be used to measure the fishing capacity of a vessel. Although the new MAGPs allow for the consideration of some characteristics of the fleet other than tonnage and engine capacity the strategy is still at an elementary level. Therefore, the aid for the modernization of vessels as well as the aid given for the temporary withdrawal of fishing vessels become mainly welfare measures.

Overall, The CFP has failed to fulfil its objectives, the main reason being that it has been based largely upon restrictions on landings without a commensurate restriction of fishing effort and without sufficiently effective enforcement and control. The result has been increased effort on the stocks, catches in excess of quotas, evasions of the technical conservation measures designed to protect the stocks and increased discarding of catches at sea.

Moreover, it has become clear that the extensive compartmentalisation of the CFP into clear-cut and distinct policies has not solved the acute problems of the fishing industry. On the contrary, the particular policies within the CFP which aim to prescribe remedies for specific problems have failed or have ambiguous results, since they are not based on a central management system which has been robustly founded. Such specific policies have failed to act synergistically with each other although they aim at the same objectives (article 39 of the Rome Treaty). In addition, the lack of coherence between the policies has inevitably produced an unbalanced treatment of biological and socioeconomic parameters among the different measures.

Another important element, stressed by the case of the Mediterranean, is that the E.C. failed to consider individual characteristics when formulating regulations. It should be considered that the E.C. is not dealing with peculiarities derived from state frameworks, but with different forms of production associated with different geographical backgrounds.

As a conclusion a comprehensive management system with enough flexibility to incorporate individual characteristics, which would be the basis for the resolution of the particular issues faced by the fishing industry, is required. The answer could be a restricted access system, which will be analyzed in the next section.

Restricted access: A proposed licensing scheme justified by a theoretical bioeconomic model

A vigorous system for maintaining the resource, which will be the backbone of the whole policy for the fishing industry, cannot be accomplished when based solely on inadequate scientific advice. After all, scientists have always been aware of the limitations of their methods. In our opinion, the inadequacy of the decision-making process of the Community is greatly to be attributed to the incoherence of the policies and the insufficient attention paid to balancing economic, social and biological parameters. Although the Community obtains scientific advice from a variety of bodies that are directly (e.g. STCF and recently STECF) or indirectly (e.g. ICES) attached to the Commission, their advice (which already incorporates uncertainty) is usually altered in the political compromises in the Council of Ministers.

It has been argued in the first chapter that in an open-access fishery fishermen disregard the future of the resource and ignore the externalities they generate by expanding their fishing effort until total revenue equals total cost. Therefore, they reduce the marginal productivity of the fishing industry as a whole. In order to impute a price and internalise these hidden costs there are currently three main approaches:

1   The privatization of the resource e.g. by allocation of individual transferable quotas (ITQs) (Clark and Major, 1988; Gean and Nayar, 1988; Kennedy, 1991)

2   Government intervention through regulation of the size and composition of the catch using a variety of policy instruments that aim for example to reduce fishing effort (e.g. TACs and quotas, gear restrictions, closed seasons etc.). Another government measure could be the imposition of an optimizing tax at a price which equates the externalities involved (Clark, 1990). A usual, though theoretical, approach is to capture the dissipated resource rent by imposing a per unit tax on total individual effort equal to the difference between the industry's average and marginal revenues at industry's optimal point of effort. This approach, although efficient in theory, is currently politically unacceptable and unlikely to be welcomed in practice by fishermen. For this reason, regulation by taxation has never

been seriously considered as a practical approach for fisheries management neither in the short nor in the long term (Anderson, 1986; Clark, 1990; Pearse, 1980).

3   The community-based management of marine resources (Berkes, 1989; Smith and Berkes, 1991). In Berkes (1989), it is argued that since market solutions cannot work if the open-access condition persists, the potential role of the traditional communal-property systems has been underestimated in fisheries management. However, there are no economic data to support this remark. Since the structure of this system do not allow for extensive planning, economic efficiency under a community-property system is not to be justified as a generalised rule. In reality such systems are viable under certain socio-anthropological circumstances which are also supported by peculiarities (biological and ecological) of certain fisheries in the world that are fished by local populations. For example, Ruddle (1989) describes such a communal based system of fisheries management in certain Japanese coastal areas where people living in villages comply more to customary procedures and community rules of conduct rather than to modern legislation rules. In other cases, for example for Lobster fisheries in Mexico (Quintana Roo) (Miller, 1989) and in the U.S.A. (Maine) (Acheson, 1989), a community-property system can be sustained mainly because of the strong territorial nature of the fishery. Under similar circumstances, it is considered that regional and local levels of government should be involved in the management of the resource (Coull, 1993).

It is of vital importance for the industry to abide with measures that promote efficiency. Inefficiency not only raises the costs of individual fishermen but also in the long term raises the price of fish and consequently weakens the competitiveness of the industry, while economic rent has been long before dissipated. As opposed to the situation created by the current CFP, a more rational and potentially more efficient proposal for fisheries management would be a combination of the above three approaches. This would enable the managers to tackle a variety of specific problems without losing their ability to balance the multi parametric nature of the fishing industry in the E.C.. In general, the solution could be the implementation of property rights for fishing the resource through a system of licences that would incorporate some other measures as well. This system can be structured as follows:

1   A number of licence units should be established. Each of them will determine a small percentage of quotas for certain fish stocks for one or more particular species (depending on geographical characteristics i.e.

migratory behaviour or regional mixed type fisheries), where information is available. Quotas will be allocated by a centralized system under the auspices of the Community.

2   Each licence will define the fishing method for these stocks. For example, each licence will be restricted to certain vessels according to their tonnage capacity and their engine power. It would also set particular gear restrictions (eg. mesh size, detection methods used) and time available for fishing.

3   Licences will in the first instance be bought by E.C. fishermen from the E.C. or they can simply be auctioned. After that each E.C. fisherman will have the right to buy or sell the licences in order to obtain the best catch revenues for his capital investment.

A string of questions may arise: for example what will be the price of each licence; why licences have to incorporate gear restrictions; why for certain types of vessels; why for certain fisheries. In order to explore the possible outcomes of such a system and reveal the rationale and efficiency of this proposed scheme, I develop a mathematical bioeconomic model which aims to show the effects of my proposed scheme on the fisheries.

Let, s, be a variable that represents the size of an individual fish (e.g. in cm) in a multicohort fishery and let $\varphi(s)$ denote the size distribution of this particular fish stock, in the sense that

$$\int_0^\Sigma \varphi(s)ds = 1.$$

The above proposed scheme of licences would permit a vessel to harvest at a rate $h \leq H$, where H denotes the permitted harvest that is specified by the accumulation of property rights in terms of gathered licence units, for example, by a fisherman who owns a vessel. In addition, the above proposed measure of licensing will incorporate gear restrictions (e.g. mesh size). Therefore a vessel operating under this scheme has to catch fish of size $\sigma \leq s \leq \Sigma$, where σ is the smallest size allowed by the licence in this multicohort fishery. Consequently, harvest is described by the following equation,

$$h = q(x)Ex \int_\sigma^\Sigma \varphi(s)ds \qquad \text{that is based on the production}$$

237

function of the fishing industry (see 1.2.1) where x denotes the size of the resource population. The catchability coefficient, q, is treated here in its most general form as a function of the current stock size x and it is assumed that $q'(x) > 0$ (Clark, 1985). In other words, the spatial concentration profile of fish stocks is not homogeneous as the constant catchability coefficient in the Schaefer model implies. Therefore, as is naturally expected a given fleet will not encounter a uniform spatial density of stocks at the same time. Furthermore, such catchability allows for the common behavioural pattern of fishermen, that is to exploit first the currently most productive fishing areas (Sampson, 1991). A sustainable aggregate quantity for the various economic inputs such as labour plus capital services devoted to harvesting, is referred to as "fishing effort", E and it is assumed continuously variable over time.

Assume that each licence unit can be bought or sold at a price l. In addition the price received for fish is size dependent $p = p(s)$ and is an increasing function of size, s. This principle is generally applied in the market for fish caught for human consumption. The net revenue flow, π, must include the opportunity cost for the licence $L = lh$ and will be equal to:

$$\pi = p(s)h(\cdot) - L(l,h) - c(E), \text{ thus}$$

$$\pi = q(x)Ex \int_{\sigma}^{\Sigma} p(s)\varphi(s)ds - lq(x)Ex \int_{\sigma}^{\Sigma} \varphi(s)ds - c(E),$$

where c(E) denotes the variable costs of effort. It is assumed that fixed costs (non-malleable capital), such as the price of the vessel, gear, port facilities, are not taken into consideration as long as a fisherman is actively in the industry and seeks to optimize his fishing effort.

Each fisherman under such a scheme wishes to choose his optimal level of effort to catch the permitted size of fish $\sigma \leq s \leq \Sigma$. As the price of fish depends on size, there must be an optimal size of fish that will give an optimal return to the fisherman.

The optimization model would be the maximization of the net economic returns for a certain period of time, $0 \leq t \leq T$, or as long as the same quotas are applied. That is,

$$\text{maximize:} \int_{0}^{T} e^{-\delta t} \pi(\cdot) \, dt$$

subject to constraint: $\dfrac{dx}{dt} = \dot{x} = F(x) - q(x)Ex \int_{\sigma}^{\Sigma}\varphi(s)ds$, x_0 given,

where the above constraint is based on Schaefer's logistic population dynamic model where the change in fish population is represented by the difference between the F(x) function that defines the net natural growth rate of the fish population minus the harvest rate, h.

By the use of optimal control theory, the current value Hamiltonian for this problem is:

$$\tilde{H} = \pi(\cdot) + \mu\left[F(x) - qEx \int_{\sigma}^{\Sigma}\varphi(s)ds \right]$$

where μ, is called the current-value shadow price of effort that represents the costs of externalities (see first chapter for more details), for example, the interspecies externalities which arise when the harvest of one species affects the abundance of other species, or "crowding externalities" that arise from the congestion of boats and gear on the fishing grounds. Thus, when one vessel exerts its fishing effort it imposes costs on other vessels because it reduces the fish stock and thereby increases the harvest costs for all the other vessels. From the Maximum Principle the optimal level of effort is given by:

$$\dfrac{\partial \tilde{H}}{\partial E} = 0 \text{ that is } c'(E) = qx \int_{\sigma}^{\Sigma}\big(p(s) - 1 - \mu\big)\varphi(s)ds \qquad \text{(EQ. 1)},$$

and the optimal size limit of fish, σ, must satisfy $\dfrac{\partial \tilde{H}}{\partial \sigma} = 0.$

That implies $\qquad p(\sigma) = 1 + \mu \qquad$ (EQ. 2)

Alternatively, maximization of $\tilde{H}$ with respect to s requires the maximization of

$$\int_{\sigma}^{\Sigma}\big(p(s) - 1 - \mu\big)\varphi(s)ds \qquad \text{(EQ. 3)}$$

In order to set eq. 1 in a realistic context i.e. in a market economy, $c'(E)$ represents the Marginal Cost of fishing effort and requires a profit rather than a loss. This means from (EQ. 3) that this inequality, $p(s) > 1 + \mu$, must hold for all fish sizes $\sigma < s \leq \Sigma$. The minimum size of fish that determines the marginal economic viability for the fisherman (ie. taking account of the opportunity of selling licences rather than catching low-priced small-sized fish) will be $p(\sigma) = 1 + \mu$. This means that the opportunity of selling the licence at price, 1, plus the current shadow price, μ, must at least be equal to the price $p(\sigma)$ obtained by selling the smallest permitted size of fish.

Equations 2 and 3 imply that fishermen must be willing to accept selective gear measures (e.g. mesh sizes) that efficiently allow them to catch fish bigger or at least equal to size σ, in order to benefit from fishing. Therefore, in order to correct deficiencies in the price mechanism by the licensing method, in an economic cost-benefit analysis the price of a licence must incorporate a stream of externalities, for example environmental degradation from overfishing, gear, pollution and crowding externalities as well as the administrative costs of managing the resource. This can be obtained by setting the new price of a licence equal to $1^* = 1 + \mu$. Therefore, if we want to establish the optimal price of a licence, 1^*, that will be sold to fishermen for the first time or is to be the lowest permitted bid in an auction, 1 will represent the costs of management of the resource and μ will retain the same notion i.e. the current value shadow price of the resource. Thus, the fishing industry will operate at a bionomic equilibrium. This does not necessarily imply that the economic rent, or net revenue, will be completely appropriated. In practice, although the least efficient vessels will earn zero rent, there will be more efficient vessels that will earn positive rent in this equilibrium. Such a management system, where the private costs of fishing are at a level similar to that of social costs, could provide the solution to many of the problems mentioned above.

The model implies that the licence scheme applied in a multicohort fishery has to include explicitly certain TACs, quotas and fish size restrictions. Although such measures exist under the CFP their role remains completely conservationist. Under the proposed licensing system the effects of such measures on the economy are also considered (externalities generated will be consolidated in the price of the licence). The licences will be managed in such a way that their application will generate economic incentives and consequently additional income for fishermen who will have the option to retain or sell licences. On the other hand, the proposed method of licensing will create a feedback mechanism by means of the "conservation measures" (i.e. TACs, fish size restrictions etc), so that the price of a licence can be adjusted adaptively to the reality of biologic and economic forces in a market economy.

Furthermore, the above model is based on the argument that solely restricting access by means of licensing fishermen is not adequate either for conservation or economic rationality in the fishery. Overcapacity might increase, for the remaining fishermen could increase their fishing effort, for example by increasing fishing time, vessel capacity or fishing gear. Fishing licences should also contain other measures such as TACs and quotas, vessel size, gear restrictions, time available for fishing etc that control the fishing effort of the licensed fishermen. This is not only a theoretical argument; empirical evidence readily supports it. For example, in the British Columbia salmon fishery a government scheme of redundancy payments reduced licensed fishermen during 1968 to 1974. However, total fishing capacity was not reduced but increased. This was mainly due to the remaining fishermen replacing gillnets with purse seines and small seiners were replaced by larger more powerful vessels (Fraser, 1978; Rettig and Ginter, 1978). In addition similar incidents have been reported in various Australian fisheries (Hilborn and Walters, 1992; Meany, 1978).

Although the proposed system of restricted access constitutes a drastic change to the existing state of open access in the industry, it is claimed to be a more politically favourable system than most of the other measures that aim to reduce fishing capacity, for example taxation. This is proved by the numerous applications of similar systems in world fisheries based on imputing a price to the resource by restricting access, such as Individual Transferable Quotas, or fishing licences. Their application in many countries, for example Canada, Iceland, Netherlands, New Zealand, Australia (Clark and Major, 1988; Gean and Nayar, 1988; Kennedy, 1991; McKellar, 1990), seems to have confirmed the expectations of economists. However, Copes (1986) sceptically examines their long-term consequences. The proposed system makes provision for long term effects by requiring that external diseconomies also be taken into consideration. Thus, it contributes to the more efficient rationalisation of the fleet and consequently to the control of the fish stocks in a more efficient way than the above mentioned systems.

Regarding the excess capacity in the fishing industry, there will inevitably be fishermen who will be made redundant by the proposed system. However, the scale of redundancy and the consequent socioeconomic effects could be attended to by the regional social policy if it is considered appropriate. For example, in order to protect the fishing industry in areas that are traditionally dependent on fishing in the E.C. (as appears in the structural regulation of 1986 [(146)]) special grants could be provided for the purchase of licences by fishermen in those areas. However, such a preferential system in these areas must be accompanied by measures to promote efficiency. For example, such rights may be accompanied by a strict, time-scheduled scheme similar to national MAGPs, which will set realistic objectives for the regional fishing industry. In the case of a failure to observe these, then the licences or a part of them could be bought by others outside the region. A possible reaction by local fishermen could be to promote joint collaboration towards these objectives and raise their efficiency in these coastal areas, thereby becoming competitive with capital intensive fishing. This could be done by forming Cooperative Fishermen's Associations. In addition, national authorities can more easily monitor and control such associations than they can individual fishermen. It can be also argued that given that fishermen who belong to such small scale cooperatives have similar fishing activities, interests and objectives (Jentoft, 1985), expert consultation and advice could be more accessible.

In welfare economics social welfare is based on both efficiency and equity. For devising the conditions to attain some optimum involves the judgement that the optimum chosen is ethically desirable. Thus, the amount of compensation or even compensation itself would be regarded relative to what society contemplates as "distributive justice" (Mueller and Wang, 1981). The acknowledgement of distributive justice in Europe has been shown in many instances (e.g. redundancy payments for industrial workers, national health, unemployment benefits, financial contributions by organizations and individuals for support of lengthy strikes, charities etc.). With regard to fishermen who will become redundant under such a scheme of licences, a compensatory payment could be met by government intervention under a buying-out excess of capacity. The whole area of income distribution and the social consequences derived by the application of regulatory measures in the fishing industry, given the job immobility of the sector, need further research since the existing CFP has not adequately considered this area (see conclusions in the structural policy chapter).

By the application of such a scheme of licences, a licence will become a capital asset of substantial value for particular areas where "good catches" are expected. Given that fishermen seek to maximize the benefits of their capital investment, they should be keen to know the state of fish stocks, since the price of the licence would reflect this. Fishermen therefore, would be more inclined to support and actively assist the monitoring system.

The implementation of CFP reforms

The application of the above generalised theoretical model for fisheries requires that the externalities in a licence scheme should be appraised and internalised in the price of the licence in order to promote efficiency in the fishing industry.

Further research needs to be done on the ways of estimating the shadow values and internalizing the externalities. This is also stressed by Tucker and McKellar (1993), who explored the estimation of "shadow prices", defined as the discounted present value of future profits to be gained from reducing the current catch of a certain aged class fish by one small unit of biomass. Their approach requires extension to the real levels of complexity, for example the inclusion of multi-fleet and multi-species fisheries and effort data which will incorporate the segmentation of input. Their work, although preliminary, could result in a valuable methodology which could be used for the estimation of the degree of structural overcapacity. Recently, economists have explored the possibility of applying capacity utilization measures in order to predict the effect of regulatory constraints i.e. quotas on the investment tendencies of the fishing industry (Segerson and Squires, 1993) in an unregulated fishery. In that way, one could predict the effect of a proposed regulation (e.g. quotas or licences in the fishing industry) before it is applied, by a methodology which uses virtual prices (i.e. those prices which would induce an unrationed market economy to behave in the same manner as when faced with given ration constraints) in order to model the effects of rationing in the fishing industry. It is based on the work of Neary and Roberts, (1980), who showed that all the properties of the rationed demand and supply functions may be expressed in terms of the properties of unrationed functions, provided that the latter are evaluated at virtual prices. Virtual prices are also functions of ration levels and can be calculated from the unconstrained demand and supply functions. More research needs to be done by researchers in order to develop such models which can be of considerable value for policy formation.

The main advantage of the proposed theoretical model is that its general form allows for a different treatment of the particular characteristics in each fishery. For example, depending on the ecological and biological characteristics of a given fish stock a different population dynamics model for $F(x)$ can be employed (e.g. Beverton-Holt type dynamic pool models) for the constraint function with only minor alterations of the generalised form given above. However, the data requirement of such a proposed licence measure is demanding and furthermore extensions to the real levels of complexity of fisheries need to be considered as happens in multi-fleet fisheries and multi-fishery fleets.

The data requirements demand sound knowledge of the population abundance and the size distribution of fish stocks. In addition, the spatial concentration profile of fish stocks must be known and then it can be incorporated in the catchability coefficient, $q(x,\rho)$ as a function of the population size x, and the

243

density ρ, of the exploited population. The requirement for such data is essential for the knowledge of ecological and biological characteristics of the exploited fish stocks, for example on how they occupy the available habitat and which biological and environmental parameters it is important to sustain for the fish stocks concerned. For example, the model requires the knowledge of interspecies relations because these could generate external disseconomies since the exploitation of one stock could possibly affect the abundance of other species. Furthermore, the relationship between recruitment and the population of fish stocks determines also the externalities that are generated by catching small-sized and young-aged fish (see also second chapter). Therefore, such characteristics of the population dynamics of fish stocks are important for the pricing of the licences proposed as well as for the setting of TACs.

Another parameter which needs a reliable definition for its incorporation into the licence scheme is fishing effort. The engine power and tonnage of the fishing vessels are not sufficient indicators of the pressure applied to the fishing stocks and more factors should be included and weighed accordingly. These parameters should include time spent in fishing, the efficiency of sophisticated detection equipment, facilities for handling the catch on board, capacity of the fish hold on board. e.t.c. The segmentation of the fleet under the licence scheme is necessary, since it is directly related to the pressure exerted on fish stocks and consequently for externalities that arise from the different kind of fishing activities of the fleet. The influence of the parameters of the production function for effort on the social net benefits of licence limitation has been recently investigated by economists (Anderson, 1985; Campbell and Lindner, 1990). Anderson (1985) has demonstrated that a licence scheme which aims at restricted inputs can generate rents, because the reduction in the total amount of fishing effort will cause a shift of resources to higher value uses elsewhere. Campbell and Lindner (1990) have investigated the subject further and they used the marginal cost function for effort, which is derived from the production function, as a means to evaluate the performance of an optimal licence limitation programme. They postulate that a limitation programme is likely to produce net social benefits as long as non-restricted inputs cannot easily be substituted for restricted inputs of effort and/or as long as restricted inputs enter the production function with a significant proportion of total cost. However, if their findings are to be used for efficient decision-making further investigation is required since their work is based on the oversimplistic static Schaefer model and in linearity assumptions without considering the effects of discounting.

The licence scheme should incorporate comprehensive information on fishing effort and stock abundance. The scheme appears to be very restrictive, since there will be a limited number of licences sold, but it is the more aggressive management that produces the greater economic efficiency. However, the social and political environment affects the success of limited entry plans. Effective management is very difficult when the political environment insists upon government guarantees of fishermen's incomes. Subsidies make the necessary exit from the fishery virtually impossible to obtain.

Although the political will for introducing a system of licences attached to each vessel is expressed in the recent 1992 regulation [155], such a system requires the establishment of additional regulations in order to operate efficiently. The data requirements for an efficient scheme of licences exceed by far the data which are currently available. Therefore, inevitably such a scheme is not to be applied in the short term. It is mainly data gathering that postpones the application of this scheme to the long term perspective since the current administrative burden would be reduced. For example, under the new system there will be no place for the cumbersome and complicated market intervention scheme. Moreover, the role of producers' organizations would be diminished since the objective for which the producers' organizations were established i.e. "taking measures which will ensure that fishing is carried out along rational lines and that conditions for the sale of their products are improved" (article 4(1) of the 1992 marketing regulation [72] would be fulfilled. Therefore, the budgetary provisions for such measures under the existing CFP can be allocated to the establishment of a comprehensive monitoring and surveillance system as follows.

Under the 1992 regulation that establishes a Community system for fisheries [155] the Community moves towards the unrestrained access to E.C. waters by Member State fleets after the year 2002. Therefore, a centralised administrative system for fishing licences would be more appropriate. The establishment of such an administration is not expected to cause considerable inconvenience and be a substantial burden on the Commission. There is already a unit under the Commission's responsibility that is responsible for issuing licences for Community vessels fishing in third country waters. The Commission handles efficiently this responsibility and therefore, an expansion of this unit would provide the administration for a licence system for fishing vessels. There is no reason why such a centralized unit will fail to meet the administrative requirements for a licence scheme. In all Member States there are similar units that authorise the licences of numerous vehicles.

In November 1993, the Commission presented a proposal [191], which resulted in a Council regulation establishing a Community system of fishing licences [192], as was required by the 1992 regulation [155]. This new regulation will fully apply from 1996 and will provide for the establishment of fishing licences. It also laid down the rules on the minimum information to be contained in the licences, for example main type of gear, engine power, length and tonnage. As expected, the licences will be issued by the flag Member State, whereas the Commission will be responsible for issuing the licences to third countries. Although the above licensing system will undoubtedly improve the monitoring of fishing activities, it fails to incorporate all the factors that affect fishing effort, e.g. the use of detection electronic equipment. Moreover, it is not a rent generating measure, as would be the licence scheme developed above, the revenues from which could be used to tackle the problem of overcapacity.

The importance of a coherent monitoring, surveillance and enforcement system for short term CFP reforms

The lack of information or complete data frequently represents the main obstacle to analysing particular policies under the CFP The measures under the previously examined fisheries regulations of the CFP have always been designed to exert control over specific aspects of the CFP without being particularly concerned about their effects on other aspects. For example, the Community grants assistance for construction and modernization projects for fishing vessels without taking into account their compliance with conservation measures. The bulk of the measures have been designed without balancing the importance of biological, economic and social parameters and this has led to the compartmentalization of the CFP into distinct and separate components. As a result measures are not synergistic and the CFP, although it aims to be the most integrated policy in the Community, lacks coherence between its various components. For example, there is not a single measure in market organization that pertains to conservation policy or to structural policy. The net result is that fisheries management in the Community is almost left solely to the establishment of TACs and quotas. Given that, a logical assumption would be that priority has been given to the biological approach. However, this is far from being a solid argument, since TACs are finally settled by politicians in the Council of Ministers.

In this situation, which is determined by a lack of consistency in the different components of the CFP, the monitoring policy of the Community would be surprising if it were coherent. Although the Community established in 1987 basic monitoring legislation [177], which amends that of 1982 [178], the system introduced is restricted to monitoring compliance with the rules that concern the conservation of the resource. In addition, according to the "a la mode" principle of "subsidiarity", the 1987 regulation established the rule that each Member

State is primarily responsible for monitoring fishing activities in its own territory. It was late 1989 [179] when the Community established a derogation from this kind of "subsidiarity" by strengthening a joint monitoring scheme, supervised both by the Member State and the Community. Under this new monitoring scheme the Community increased the aid and the contribution given to Member States, within the range of 35-50%, for naval and aerial surveillance and for modernizing data-processing equipment (figure 8.1). A new Council regulation, which aims to combine the previous measures and further ameliorate the establishment of a control system applicable to the CFP, was presented very recently [190]. It came into force on the 1st of January 1994 and it represents a more comprehensive attempt at monitoring all fishing activities. It introduces the establishment of a computerized database which will improve the cross-checking and verification of data by the Member States and the Community and elucidates the responsibilities of the parties concerned for the efficient monitoring of the fisheries.

However, the accomplishments of the Community's monitoring policy are largely disputable. Considerable gaps in recording catches and notifying them to the Commission are evident in the statistical tables for fisheries in most Member States [148]. In addition, Mr. Banks, the chief executive of the National Federation of Fishermen's Organisation of the U.K. in 1992, has accused Spanish and French fishermen of systematically disregarding conservation regulations. The French and Spanish authorities do not even monitor these infringements [49]. For example, Community inspectors who attempted to check the fish markets in the Spanish ports of La Caruna and Vigo, which are flooded with illegal under-sized fish, were not allowed to do so. He concludes that there seems to be a conspiracy between the local fishermen and the Spanish authorities. Such incidents, coupled with the inability of the Community's inspectorate to carry out independent spot checks, make the reliability of the existing data disputable, a problem which is unlikely to be resolved by the 1993 regulation. Furthermore, the Commission has no direct access to the information included in vessel logbooks and, therefore, the reliability of data does not only depend on fishermen's honesty but also on the commitment and the firmness of the political authorities in the Member States which should allow such access.

In addition, increases in expenditure (which increased in 1993 [181], figure 8.1), or a more rational reallocation of the existing budget alone is unlikely to result in a better surveillance and comprehensive monitoring and control system. For example, the Community's expenditure for the construction of fishing vessels, which is not only contradictory with the aims of the MAGPs but also generates overcapacity and increases the inefficiency of the fishing

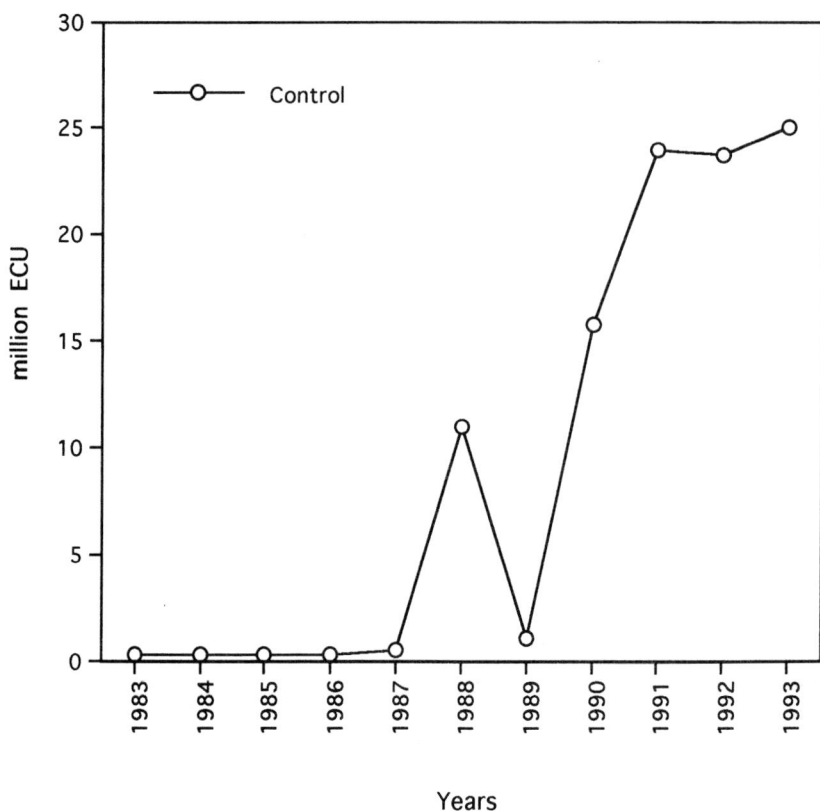

note: Data1983-1985 for 10 Member States, data 1986-
1993 for 12 Member States

Figure 8.1 Community expenditure on monitoring
and surveillance during 1983 to 1993

source: Data from SEC(91) 2288 (18.12.1991) and General Budget of
E.C. in O.J. (8.2.1993) L31 and O.J. (15.2.1993) C330

248

industry, could be more rationally allocated to monitoring policy. In 1991 the Community expenditure for shipbuilding was 47,24 million ECUs which is about double the amount for the monitoring and surveillance policy. However, such an increase could be meaningless since in 1991 only 39% of the Community's expenditure for monitoring policy were utilised. Therefore, the lack of commitment and firmness of the political authorities in the Member States rather than a budgetary issue seems to be the major obstacle to a better comprehensive monitoring policy for the Community's fishing industry.

It is a fair assumption that the inefficiency of the monitoring system generates ineffectiveness in the enforcement of the rules. However, even if there were a satisfactory monitoring system in force and the political will in the authorities to trigger sanction proceedings for the offenders, the inadequacy of the existing rules would also make the enforcement uncertain. It is not only the complexity of the CFP legislation but also the national legislation of Member States that enhances this complexity by offering a number of derogations to the existing rules for fishing at particular times of the year and in certain areas with specific gear (e.g. seasonally closed areas without the overall banning of fishing). As a result a possible offence is difficult to prove. Even in cases where fines have been imposed on fishermen, there is a general tendency in the Community for fishermen to aid each other by jointly paying the fines. Moreover, fines could work as an additional incentive to catch more fish in order to offset the additional cost.

An important component necessary to substantiate the monitoring system is the determination of national governments. However, the existing fisheries management system (characterized by TACs and quotas and the granting of financial aid and subsidies in the structural and marketing policies) provide little incentive to fishermen and governments to empower a reliable and comprehensive monitoring system for fishing activities. Therefore, under the existing CFP a more centralised monitoring system is needed which would invest in Community inspectors more powers than those of the national authorities. However, such a system is unlikely to be applied since again fishermen and Member States' governments are not committed to this policy.

An improvement in the traditional monitoring of fishing operations could be secured by using the modern techniques of radionavigation based on recent developments in satellites. The Commission launched a pilot experimental project called MONICAP [180] for the Portuguese fishing fleet in March 1989. A satellite system is capable of identifying and locating fishing vessels which carry a device called a "transponder" which is interrogated by a satellite and the information is transmitted to earth stations which are already linked to the existing surveillance networks. The degree of accuracy for the positioning of a vessel by a satellite used by civilian users is around 100m. By determining the location of a vessel at a time its velocity can be estimated, which for some types of fishing methods (trawling, dredging, drift nets) is a good indication of fishing effort. Although such a system has many merits the price of a radionavigation receiver which is between 1,500 to 2,000 ECUs may be excessive for small scale of fishermen. However, given the importance of a monitoring system and the Community's low expenditure on monitoring policy (only 4% of the total budget for fisheries in 1993, figure 5.2, page 119), the Community requires to increase its expenditure in this area. In the latest Council regulation of 1993 [190], the Community renewed its support for continuous position-monitoring pilot projects until 1995 and is committed to decide by 1996 whether to adopt a uniform land- or satellite-based monitoring system.

Although the latest regulation [190] represents an improvement in effective control for all fishing activities, it underlines the inefficiency of the existing management system. To be effective the existing CFP obviously requires extensive policing methods. Such measures are unlikely to be welcomed by fishermen and national authorities, who would rather continue with their present practices of inaccurate figures which understate landings.

Conclusions

The fisheries management under the CFP fails to live up to the expectations of European producers and consumers, let alone the stocks. The deficiency can be traced mostly to the attempt of the managers to regulate the fishery output, rather than the input. This is essentially unsuitable given the uncertainty and variability of the biological advice. The situation is intensified with the combination of political compromises, inflexibility and neglect of economic principles. Therefore, a new comprehensive approach is called for.

A licence system is proposed as an alternative to the current fisheries policy. In that way a limited number of licences will be released on the market. Licences will be issued for different types of vessels, stating the gear to be used, and for a certain percentage of quotas of one or more species, where information is available. The licences will be freely transferable.

The licence system represents a highly restrictive measure, which is hoped to benefit the fisheries sector by reducing overcapacity and leading to a robust monitoring system, which in turn would result in the long-term recovery of fish stocks which are currently overexploited. In addition, regional features would be able to be incorporated in the licence characteristics.

Further research needs to be done by biologists in order to develop ways of predicting more accurately the stock situations. Economists are also called upon to evaluate the economic parameters which affect the fishing industry and incorporate them in the licence scheme.

However, it will take more than economic experts for the scheme to make a difference to the achievement of efficiency. It is decisive politicians, willing to accept the political risks, that can implement and enforce such a policy and see it being successful. It is unlikely that a gradual approach will lead to considerable achievement, since pressure from constituencies will tend to reverse the restrictions imposed. Therefore, political will represents an important, if not the most important, component for the success of the proposed schemes.

Notes

1 Reg. 170/83 O.J. (1983) L24/1, article 12.

2 Dec. 572/79 O.J. (1979) L156/29.

3 Annex I, XV O.J. (1985) L302/241.

4 Dec. 128/71 O.J. (1971) L68/18.

5 Dec. 429/73 O.J. (1973) L355/27 article 2.

6 Dec. 441/74 O.J. (1974) L243/19.

7 Dec. 252/68 O.J. (1968) Sup. Ed. (1), p.125.

8 Dec. 441/74 O.J. (1974) L243/19, articles 2 and 3(1)(a).

9 E.C. Commission (1966), COM (66), 250 final, 22-6-1966.

10 Reg. 2141/70 O.J. (1970) L236/1.

11 Reg. 2142/70 O.J. (1970) L236/5.

12 E.C. Commission (1976), COM (1-23), 500 final, 23-10-76.

13 E.C. Commission (1980), COM, 575 final, 1-10-1980.

14 SEC (1975), 4503 final, (9-10), 22-12-1975.

15 O.J. (1981) C105/1, 1-2-1981.

16 O.J. (1980) C158/20, 27-6-1980.

17 Judgement of the European Court of Justice of 16 June 1987, case 46/86, par. 17.

18 Reg. 2142/70 O.J. (1970) Supl. Ed. 3, p.707, articles 28-30.

19 Reg. 3796/81 O.J. (1981) L379/1, articles 32-34.

20 Reg. 170/83 O.J. (1983) L/24/1, articles 13-15, 27-1-1983

21 Reg. 2141/70 O.J. (27.10.1970) L236/1, article 11.

22 O.J. (1980) C342/1, articles 86,87.

23 E.C. Commission (1980), COM, 338 final, 12-6-1980.

24 SEC (1981), 105 final, 21-1-1981.

25 O.J. (1982) C22/2.

26 Reg. 170/83 O.J. (1983) L/24/1, articles 6,8, 27-1-1983.

27 Reg. 170/83 O.J. (1983) L/24/1, 27-1-1983.
28 Reg. 172/83 O.J. (1983) L/24/30, 27-1-1983.
29 Reg. 171/83 O.J. (1983) L/24/14, 27-1-1983.
30 O.J. (1983) C28/1, 3-2-1983.
31 Reg. 2166/83 O.J. (1983) L206/71, 29-7-1983.
32 O.J. (1982) C228/1, 1-10-1982.
33 Reg. 2527/80 O.J. (1980) L258/1.
34 Reg. 2108/84 O.J. (1984) L194/22, and Reg. 3/84 O.J. (1984) L318/23.
35 Reg. 3421/84 O.J. (1984) L316/34.
36 Commission's answer to written question 2037/85, O.J. (1986) C182/10.
37 Reg. 1866/86 O.J. (1986) L162/1.
38 Reg. 3499/91 O.J. (1991) L331/1.
39 17th, 18th and 19th General Reports on the Activities of the European Communities pp.195, 186 and 228 respectively.
40 Reg. 3094/86 O.J. (1986) L288/1.
41 Reg. 350/77 O.J. (1977) L48/28
42 Judgements of the European Court of Justice, case 141/78, 4 October 1979, O.J. (1979) C280/5, and case 32/79, 10 July 1980, O.J. (1980) C199/6
43 Act of Accession of the Kingdom of Spain and the Portuguese Republic O.J. (1985) L302/1.
44 Reg. 3531/85 O.J. (1985) L336/20
45 For vessels of the 10 operating in Spanish and Portuguese waters Reg. 3716/85 and Reg. 3719/85 O.J. (1985) L360/7 and 26. For Spanish vessels in Portuguese waters and Portuguese vessels in Spanish waters Reg. 3718/85 and 3718/85 O.J. (1985) L360/14 and 20. For Spanish vessels in waters of the 10 Reg. 3531/85 O.J. (1985) L336/20. For Portuguese vessels in the waters of the 10 Reg. 3715/85 O.J. (1985) 360/1.
46 Data published in SEC(92) 2340 final (23.12.1992) "Report (1992) by the Commission to the Council and Parliament on the Application of the Act of Accession of Spain and Portugal in the fisheries sector", Annex II.
47 Reg. 1381/87 O.J. (1987) L132/9
48 Reg. 101/76 O.J. (1976) L20/19
49 House of Lords, Select Committee on the E.C., "Review of the Common Fisheries Policy", session 1992-93 (2nd report)
50 European Court Judgements on 14 Dec. 1989 in Cases 3/87 and 216/87.
51 European Court judgements on 25 July 1991 in cases 221/89 and 4 October 1991 in cases 246/89.
52 Fishing News,1991, number 4072, Lowestoft.

53 European Court judgement of 16.6.1987 Case 53/86, Officier van Justie for the District of Zwolle v. Albert Romhes, O.J. (1987) C183/1.

54 Commission of the E.C., "Report 1991 from the Commission to the Council and the European Parliament on the Fisheries Policy", SEC(91) 2288 final, 18.12.1991.

55 Reg. 345/92 O.J. (1992) L42/15.

56 Reg. 3655/91 O.J. (1991) L348/14.

57 Reg. 100/76 O.J. (1976) L20/1.

58 Reg. 3796/81 O.J. (1981) L379/1.

59 Reg. 1865/82 O.J. (1982) L206/1.

60 Reg. 827/68 O.J. (1968) L151/16.

61 Reg. 2136/89 O.J. (1989) L212/79.

62 Reg. 1536/92 O.J. (1992) L163/1.

63 Council Directive of 22 July 1991, 91/493/EEC, O.J.(1991) L268/15, 24.9.91.

64 Com. Dec. 93/54/EEC, 22.12.92, O.J. (1993) L13/18.

65 Com. Dec. (3/185/EEC, O.J. (1993) L79/80.

66 Com. Dec. 93/140, O.J. (25.1.1993) L16/22 and Com. Dec. 93/140, O.J. (9.3.1993) L56/42.

67 Reg. 3703/85, O.J. (1985) L351/63.

68 Court of Auditors, "Special Report on the Common Organisation of the Market in Fishery products", O.J. (1985) C339/1-6.

69 Reg. 2062/80 O.J. (1980) L200/82.

70 Reg. 1995/84 O.J. (1984) L186/23.

71   Reg. 3940/87 O.J. (1987) L373/6.

72 Reg. 3759/92 O.J. (17.12.1992) L388/1.

73 Reg. 3687/91 O.J. (23.12.1991) L354/1.

74 Reg. 3140/82 O.J. (22.11.1982) L331/7, Reg. 1452/83 O.J. (1983) L149/5 and Reg. 671/84 O.J. (15.3.1984) L73/28.

75 For the first, second and third year the aid must not exceed 3%, 2% and 1% respectively of the value of production marketed under the auspices of the producers' organizations, nor must it exceed 60%, 40% and 20% of the organizations' administrative expenses in these years respectively. The above mentioned value of production marketed is established on a standard basis each year and represents the members' average marketed production during the three calendar years preceding the period for which aid is requested. Article 7, paragraphs 2,3 of the 1992 market regulation *op. cit.* in number 72.

76 Data published in the "report 1991 From the Commission to the Council and the European Parliament on the Common Fisheries Policy", SEC(91) 228 final, (18.12.1991), Annex XI-1.

77 Data publised in the "Final Adoption of the General Budget of the E.C. for the Financial Year 1993", O.J. (8.2.1993) L31/692 and 695.

78 The two schemes are identical since for the products listed in Annex I E their selling price is fixed by the Community with the same conditions as the withdrawal price. Article 13, *op. cit.* number 72.

79 These products are referred to in Annex II of the 1992 marketing regulation, op. cit. number 72 and are halibut, hake, sea bream, shrimps, cuttlefish, squid and octopus.

80 Opinion of the Economic and Social Committee, 92/C313/10, O.J. (30.11.1992) C313/22.

81 Reg. 1495/89 O.J. (1.6.1989) L148/1.

82 Reg. 3901/92 O.J. (23.12.92) L392/29.

83 Reg. 3902/92 O.J. (23.12.1992) L392/35.

84 Commission of the E.C., European File 3/91, "The Common Fisheries Policy", 1991.

85 Data published in *op. cit.* number 54.

86 Young T. (1984) "Market Support Arrangements for Fish: the Withdrawal Price Scheme", Sea Fish Industry Authority Occasional Paper Series No. 1.

87 EUROPECHE, (Brussels,15.10.1991), "EUROPECHE's position on the development and future of the Common Fisheries Policy", EP (91)22 rev.

88 Reg. 802/68 O.J. (1968) S. Edn. (I) p.165.

89 European Court judgements on Case 24/68, *Commission v. Italy* (1969) E.C.R. 193 and in C.M.L.R. (1971) 611 also in *United Foods and Van den Abeele v. Belgian State*, Case 132/80 (1981) E.C.R. 995 and in C.M.L.R. (1982) 1, 273.

90 European Court judgement on Case 132/80, *United Foods and Van den Abeele v. Belgian State, op. cit.* in number 89.

91 O.J. (1982) C245/1.

92 European Court judgement on Case 32/79, *Commission v. United Kingdom*, E.C.R. (1980) 2403.

93 O.J. (18.4.1989), 60/C32/101, C32/8.

94 O.J. (1982) C111/9, Commission's answer to Mr. Kirk W.Q. 1600/81.

95 O.J. (21.2.1990), 90/C41/06, C41/7.

96 Third Lomé Convention, O.J. (1986) L86/1, article 130 and Annex XII.

97 O.J. (1978) L266/2.

98 O.J. (1978) L263/2.

99 O.J. (1978) L265/2.

100 O.J. (1976) L264/2.

101 O.J. (1984) L343/6.

102 Reg. 1985/74 O.J. (1974) L203/30.

103 Detailed rules of the application of this system can be found in the Reg. 848/92 O.J. (1992) L89/4.

104 Research and Documentation Paper No. 22 (1991) in the EP series "Agriculture and Fisheries", "Bilateral EEC/Third country agreements and International Conventions Involving the EEC, in the Fisheries Sector", Annex II in particular.

105 Own calculations on data published in *op. cit.* numbers 54 and 77.

106 The Council authorised the Commission to negotiate with these countries found in O.J. (21.11.1992) C304/8 for Estonia, O.J. (21.11.1992) C304/12 for Latvia and O.J. (30.12.1992) C346/5.

107 *Op. cit.* in number 77 pp. 1446-1452.

108 UN resolution No. 44/225 December 1989.

109 Dec. 91/309/EEC, O.J. (28.6.1991) L166/84.

110 Details on this crisis can be found in the Research and Documentation Paper No. 21 (1991) in the EP series "Agriculture and Forestry", titled "Fisheries relations between E.C. and Canada" and in "Fisheries in the GATT Uruguay Round" documentation EP (1991) 151.955.

111 These are the Commission's preparatory act COM (92) 394 final in O.J. (8.10.1992) C259/4, and Council Regulations No. 3927/92 and 3928/92 in O.J. (31.12.1992) L397/67 and in O.J. (20.12.92) L397/81 respectevely.

112 Council Decision No. 81/1053/EEC in O.J. (31.12.1981) L379/53.

113 This exchange of letters was concluded by a Council Decision 83/652/EEC in O.J. (31.12.1983) L371/34.

114 Reg. 2054/88 O.J. (12.7.1988) L181/1.

115 Reg. 721/91 O.J. (26.3.1991) L78/10.

116 Commission's proposal in O.J. (24.8.1992) C217/5.

117 Data from *op. cit.* number 104 Annex X.

118 Reg. 223/85 O.J. (1.2.1985) L29/8 and Reg. 224/85 O.J. (1.2.1985) L29/13 this regulation amended by the Reg. 4054/89 and the Council Decision of 89/650/EEC in O.J. (30.12.1989) L389/65 and /80 respectively.

119 Part-session of European Parliament in O.J. of 12 July 1990, A3-132/90.

120 More details about these agreements can be found in *op. cit.* number 104.

121 Reg. 2210/80 O.J. (29.8.1980) L226/7.

122 Joint Cases 3,4, and 6/76, Officier van Justice v. Kramer (1976) E.C.R. 1279 (1976) 2, C.M.L.R. 440.

123 U.N. Resolution (44/225) adopted in 22.12.1989, in General Assembly, Distr. General A/Res/44/225, (30.7.1990).

124 FAO fisheries report 1991, No. 449, doc.FIPL/R449.

125 Review of Fisheries in OECD member countries, OECD Paris 1993.

126 DG De La Peche XIV-A-3, Situation au 30-6-93. Own calculations.

127 Commission to the Council COM(87) 153 final.

128 Data for 1988 from U.N. Statistical Yearbook (37th issue) 1992.

129 On the Conclusion of the Fourth ACP-EEC Convention (Lomé IV Convention) in O.J. (17.8.1991) L229/1.

130 COM(93) 214 in O.J. (18.6.1993) c167/1.

131 Conclusion of agreement between EEC-Seychelles in O.J. (20.5.1993) L124/5.

132 Conclusion of agreement between EEC-Seychellles in O.J. (6.11.1990) L306/1.

133 Conclusion of agreement between EEC-Mauritania in O.J. (31.12.1987) L388/1.

134 Conclusion of agreement between EEC-Mauritania in O.J. (10.5.1991) L117/1.

135 Reg. 2141/70 *op. cit.* in number 21.

136 Commission's "Report on the Situation in the Fisheries Sector of EEC Member States and the Basic Principles for a Common Policy" in O.J. (1967) p.862.

137 Resolution of 19.12.1980 in O.J. (1980) C346/112.

138 Regulation 17/64 in O.J. S. Edn 1963-64, p.103.

139 There is an extensive nomeclature of fishery products which is always updated. The most recent one is not yet published in the O.J. and is given by the D.G. XIV on request.

140 Slaymaker J.E. "The role of producers organizations in the fisheries management: the case of South West England" in proceedings of the IV Annual Conference of the European Association of fisheries Economists, 1992, pp.37-46.

141 According to data produced by the CRONOS data base (SEC1, collection 01) in 30 6.1993, the GDP of all factors of production in 1983 was 2387.8 bi-million ECU while in 1988 was 3608.32 bi-million ECU, that is an increase of 51.1% of the GDP in this period.

142 European Parliament, PE 151.169, May 1991

143 Reg. 2908/83 and 2909/83 in O.J. (22.10.1983) L290/1 and 9 respectively.

144 Council Directive 83/515/EEC in O.J. (22.10.1983) L290/15.

145 Reg. 2930/86 in O.J. (25.9.1986) L274/1.

146 Reg. 4028/86 in O.J. (31.12.1986) L376/7.

147 Reg. 3944/90 in O.J. (31.12.1990) L376/1.

148 Data published by D.G. XIV fisheries, XIV-A-3.(30.6.1993)

149 Regulations 1956-60/91 (8.7.1991) L181/1-114, and Commmisson's decision in O.J. (8.7.1991) L181/115.

150 Reg. 163/89 O.J. (25.1.1989) L20/5.

151 Decision 91/454/EEC, 1.8.1991.

152 Data available for Germany, Belgium, Denmark, Italy, Netherlands, U.K., Portugal and Spain. These are published in *op. cit.* number 148.

153 Decision 91/540/EEC of 14.10.1991 in O.J. L294/49.

154 Court of Auditors, Special report No 3/93.

155 Reg. 3760/92 O.J. (31.12.1992) L389/1.

156 Dec. 88/569/EEC, 28.10.1988.

157 COM reply in special report No.3/93.

158 313/09 in O.J. (8.12.1988) C88/21.

159 Report by the ACFM and the corresponding working party (1989) dated 19.11.1990 and Anon. (1990) "Scientists urge fishing ban" Fishing news, March 16.

160 EP, SEC (90) 2244, 1991.

161 Memorandum by the National Federation of Fishermen's Organisations (29.1.1992), in *op. cit.* number 49.

162 Reg. 3946/92 O.J. (31.12.1992) L401/1.

163 1992-1996 MAGPs in O.J. (31.12.1992) L401.

164 Com. Decision (91/418/EEC), O.J. (20.8.1991) L231/25.

165 Data revealed by the Economic Development & Planning Department, Grampian Regional Council.

166 Final report of the Macduff E.C. pilot scheme, dated 16 Oct. 1992.

167 PE 140.600, EN III/Q, July 1991.

168 Decision 88/569/EECof 28.10.1988, O.J. (17.11.1988) L311/40.

169 O.J. (15.12.1992) C330/107-115.

170 Reg. 3760/92, O.J. (31.12.1992) L389/1.

171 Final adoption of the general budget of the E.C. for the financial year 1993, in O.J. (8.2.1993) L 31.

172 Reg. 2794/92 O.J. (26.9.1992) L282/3.

173 Reg. 4042/89 O.J. (30.12.1989) L388/1.

174 Community Support Framework 1991-1993, Com. E.C. Doc (1992) for the U.K., Spain, Ireland, Belgium, Denmark and Greece.

175 Council Decision 88/163/EEC (2.2.1988), O.J. (18.3.1988) L72/52.

176 Holden, M. Memorandum in *op. cit.* number 49.

177 Reg. 2241/87 O.J. (29.7.1987) L207/1.

178 Reg. 2057/82 O.J. (29.7.1982) L220/1.

179 Council Decision 89/631/EEC, O.J. (14.12.1989) L364/64.

180 Commission's Decision of 2 May 1988 in O.J. (2.6.1988) L136.

181 Commisson's Decision of 3 December 1992 in O.J. L370/85.

182 21st Report of STCF, March 1992.

183 Anon. (1985) The European Community's Fishery Policy, European Documentation division IX/C/11, publisher: Office for Official Publications of the E.C., Luxembourg.

184 Data from: ICES Cooperative research reports. Reports of the ICES Advisory Committee on Fisheries Management. Volumes: No193, part 1, chapter 3.5, February 1993; No179, part 1, chapter 3.5, February 1992; No161, part 1, chapter 3.5, February 1989.

185 ICES Cooperative research reports. Reports of the ICES Advisory Committee on Fisheries Management. Volume No193, part 1, chapter 3.5, February 1993 (Copenhagen).

186 Reg. 2119/92 O.J. (29.7.1992) L213/1.

187 SEC(90) 1136 final, 10.7.1990.

188 Reg. 3499/91 O.J. (28.11.91) L331/91.

189 COM (92) 533 final, O.J. (9.1.93) C5/6

190 Reg. 2847/93 O.J. (20.10.93) L261/1.

191 COM (93) 496 final, O.J. (16.11.93) C310/13.

192 Reg. 3690/93 O.J. (31.12.93) L341/93

References

Acheson, J.M. (1989). Where Have All the Exploiters Gone? Co-Management of the Maine Lobster Industry. In F. Berkes (Eds.), *Economic Development Projects in Common Property Resources, Ecology and Community-based Sustainable Development* (pp. 199-217). London: Belhaven Press.

Anderson, L.G. (1985). Potential economic benefits fron year restrictions and licence limitation in fisheries regulation. *Land Econom., 61*, 409-418.

Anderson, L.G. (1986). *The Economics of Fisheries Management, revised and enlarged edition.* Baltimore, London: The Johns Hopkins Univ. Press.

Anon. (1965). *Survival of the Youngest Stages of Fish, and its Relation to Year-Class Strength.* (ICNAF Spec. Publ. No. 6 (363-371)). ICNAF.

Anon. (1980). *Fisheries Ecology-Some Constraints that Impede Advaces in our Understanding.* (Ocean Sciences Board No. 16). National Academy of Sciences.

Anon. (1986a). *Report of the ICES working group on the methods of fish stock assessment* No. Assess. 10. International Council for the Exploration of the Seas (Copenhagen, Denmark).

Anon. (1986b). Single European Act. *Bulletin of the E.C., 2* (Supplement).

Anon. (1987). *Report of the ICES working group on the methods of fish stock assessment* No. Assess. 24). International Council for the Exploration of the Seas (Copenhagen, Denmark).

Anon. (1988). *Report of the ICES working group on methods of fish stock assessment* No Assess 31. International Council for the Exploration of the Seas (Copenhagen, Denmark).

Anon. (1990). *Review of the State of World Fishery Resources.* (FAO Fisheries Circular No. 710). FAO (Rome).

Arrow, K.J. (1964). Optimal Capital Policy, the Cost of Capital and Myopic Decision Rules. *Annals of the Institute of Statistical Mathematics, 16*, 21-30.

Arrow, K.J. (1968). Optimal Capital Policy with Irreversible Investment. In J.N. Wolfe (Eds.), *Value, Capital and Growth, Papers in Honours of Sir John Hicks.* (pp. 1-20). Edinburgh Univ. Press.

Arrow, K.J. (1970). *Investment, the Rate of Return and Optimal Fiscal Policy.* Baltimore: J. Hopkins Press for Resources for the Future.

Baumol, W.J. (1977). *Economic Theory and Operations Analysis.* (4th ed.). Prentice-Hall.

Beamish, R.J. and McFarlane, G.A. (1983). The forgotten requirement for age validation in fisheries biology. *Trans. Am. Fish. Soc. 112*, 735-743.

Bell, F.W. (1972). Technological externalities and common property resources: an empirical study of the U.S. northern lobster fishery. *J. Pol. Econ., 80*, 148-158.

Bellman, R. (1957). *Dynamic Programming.* Princeton N.Y.: Princeton Univ. Press.

Berkes, F. (Ed.). (1989). *Common Property Resources: Ecology and Community-based sustained development.* London: Belhaven press.

Beverton, R.J.H. (1963). Maturation, Growth, and mortality of clupeid and engraulid stocks in relation to fishing. *Rapp. Procès-Verb. Cons. int. Explor. Mer., 154*, 44-67.

Beverton, R.J.H. and Holt, S.J. (1957). *On the Dynamics of Exploited Fish Populations.* (Fish. Invest. Set. No. 2(19)). Ministry of Agriculture, Fisheries and Food (London).

Beverton, R.J.H. and Holt, S.J. (1959). A review of the lifespans and mortality rates of fish in nature, and their relation to growth and other physiological factors. *CIBA Fdn. Colloq. Aging, 5*, 142-177.

Britton A. and Mayes D. (1992). *Achieving Monetary Union In Europe.* Sage Publications Ltd., London.

Bromley, D.W. (1985). Common property issues in international development. *Developments, 5(1)*, 12-15.

Burmeister, E and Dobell, A.R. (1970). *Mathematical Theories of Economic Growth.* N.Y.: Macmillan.

Caddy, J.F. and Gulland, J.A. (1983). Historical patterns of fish stocks. *Mar. Policy, 7(4)*, 267-278.

Campbell, H.F. and Lindner, R.K. (1990). The production of fishing effort and the economic performance of licence limitation programs. *Land Econom., 66(1)*, 56-66

Chapman, D.G. (1990). John Alan Gulland. *Fishery Bull. (U.S), 88(4)*, iii, iv.

Ciriacy-Wantrup, S.V. and Bishop, R.C. (1975). Common property as a conception in natural resource policy. *Natural Resources Journal, 15,* 713-727.

Clark, C.W. (1973). The economics of overexploitation. *Science, 181,* 630-634.

Clark, C.W. (1985). *Bioeconomic Modeling and Fisheries Management.* N.Y.: Wiley J. & sons.

Clark, C.W. (1990). *Mathematical Bioeconomics: the Optimal Management of Renewable Resources.* (2 ed.). N.Y., Chichester: John Wiley & sons.

Clark, C.W., Clarke, F.H. and Munro, G.R. (1979). The optimal exploitation of renewable resource stocks: problems of irreversible investment. *Econometrica, 47(1),* 25-47.

Clark, C.W., Edwards, G.W. and Friedlaender, M. (1973). Beverton-Holt model of a commercial fishery: optimal dynamics. *J. Fish. Res. Board Can., 30,* 1629-1640.

Clark, C.W. and Kirkwood, G.P. (1979). Bioeconomic model of the Gulf of Carpentaria prawn fishery. *J. Fish. Res. Board Can., 36,* 1304-1312.

Clark, C.W. and Lamberson, R.H. (1982). An economic history and analysis of pelagic whaling. *Marine Policy, 6,* 103-120.

Clark, C.W. and Munro, G.R. (1975). The economics of fishing and modern capital theory: a simplified approach. *J. Env. Econ. Man., 2,* 92-106.

Clark, C.W., Munro, G.R. and Charles, A.T. (1986). Fisheries Dynamics and Uncertainty. In A. Scott (Eds.), *Progress in Natural Resource Economics.* (pp. 99-120). Oxford: Clarendon press.

Clark, I.N. and Major, P.J. (1988). Development and implementation of N. Zealand's I.T.Q. management system. *Marine Resource Economics, 5(4),* 325-349.

Clarke, A. (1979). On living in cold water: k-strategies in antarctic benthos. *Mar. Biol., 55,* 111-119.

Cohen, J.E., Christensen, S.E. and Goodyear, C.P. (1983). A stochastic age structured model of striped bass (Morone saxalis) in Potomac River. *Can. J. Fish. Aquat. Sci., 40,* 2170-2183.

Collie, J.S. and Sissenwine, M.P. (1983). Estimating population size from relative abundance data measured with error. *Can. J. Fish. Aquat. Sci., 40,* 1871-1879.

Conrad, J.N. and Clark, C.W. (1987). *Natural Resource Economics Notes and Problems.* Cambridge Univ. Press.

Cook, R.M. and Armstrong, D.W. (1986). Stock-related effects in the recruitment of North Sea haddock and whiting. *J. Cons. int. Explor. Mer., 42,* 272-280.

Copes, P. (1970). The backward-bending supply curve of the fishing industry. *Scottish J. Pol. Econ., 17,* 69-77.

Copes, P. (1972). Factor rents, sole ownership and the optimal level of fisheries exploitation. *Manchester School of Economics and Social Studies, 40,* 145-164.

Copes, P. (1986). Critical review of the individual quota as a device in fisheries management. *Land Economics, 6(3),* 278-291.

Coull, J.R. (1993). *World Fisheries Resources.* London, N.Y.: Rontledge.

Crutchfield, J. and Zellner, A. (1962). *Economic Aspects of the Pacific Halibut Fishery.* (Industrial Research No. 11). U.S. Department of the Interior.

Csirke, J. (1988). Small Shoaling Pelagic Fish Stocks. In J.A. Gulland (Eds.), *Fish Population Dynamics.* (pp. 271-302). Chichester, N.Y.: J. Willey & Sons Ltd.

Cunningham, S., Dunn, M.R. and Whitmarsh, D. (1980). *Fisheries Economics an Introduction.* London: Mansell publishing limited.

Cushing, D.H (1975). *Marine Ecology and Fisheries.* Cambridge: Cambridge Univ. Press.

Cushing, D.H. (1988). The Study of Stock and Recruitment. In J.A. Gulland (Eds.), *Fish Population Dynamics.* (pp. 105-128). Chichester, N.Y.: J. Wiley & Sons.

Daan, N. (1980). A review of replacement of depleted stocks by other species and the mechanisms underlying such replacement. *Rapp. Procès-Verb. Cons. int. Explor. Mer., 117,* 405-421.

Daan, N. (1981). Comparison of estimates of egg production from the southern Bight cod stock from plankton surveys and from market statistics. *Rapp. Procès-Verb. Cons. int. Explor. Mer., 178,* 242-243.

Daan, N. (1987). Multispecies versus single-species assessment of North Sea fish stocks. *Can. J. Fish. Aquat. Sci., 44(Suppl. 2),* 360-370.

Deriso, R.B. (1985). Stock Assessment and New Evidence of Density Dependency. In *Fisheries Dynamics; Harvest, Management and Sampling.* (pp. 49-60). Seattle: Univ. of Washington.

Deriso, R.B., Quinn II, T.J. and Neal, P.R. (1985). Catch-age analysis with auxiliary information. *Can. J. Fish. Aquat. Sci., 42,* 815-824.

DeWilde, J.W. (1991). Review of North Sea fisheries management in the eighties. In *EAFE Annual Conference,* 3 (pp. 1-7). Dublin:

Dorfman, R. (1969). An economic interpretation of optimal control theory. *Amer. Econ. Rev., 59,* 817-831.

Emlen, J.M. (1973). *Ecology: An Evolutionary Approach.* Reading, Mass.: Addison-Wesley.

European, Parliament (Ed.). (1989). *Ten Years that Changed Europe*. Luxembourg.

Feldstein, M.S. (1964). The social time-preference rate in cost benefit analysis. *J. Econ., 74*, 360-379.

Fournier, D.A. and Archibald, C.P. (1982). A general theory for analysisng catch at age data. *Can. J. Fish. Aquat. Sci., 39*, 1195-1207.

Fournier, D.A. and Warburton, A.R. (1989). Evaluating fisheries management models by simulated adaptive control-introducing the composite model. *Can. J. Fish. Aquat. Sci., 46*, 1002-1012.

Fraser, G.A. (1978). Licence Limitation in the British Columbia Salmon Fishery. In B.R. Rettig and J.J.C. Ginter (Eds.), *Limited Entry as a Fishery Management Tool* (pp. 358-381). Washington (D.C.): Univ. Washington press.

Fraser, G.A. (1980). Licence Limitation in the British Columbia Roe Herring Fishery: an Evaluation. In N.H. Sturgess and T.F. Meany (Eds.), *Policy and Practice in Fisheries Management* (pp. 117-138). Cambera: Australian Gov. Pub. Service.

Garcia, S. (1988). Tropical Penaeid Prawns. In J.A. Gulland (Eds.), *Fish Population Dynamics.* (pp. 219-249). Chichester, N.Y.: J. Wiley & Sons Ltd.

Garrod, D.J. (1977). The North Atlantic Cod. In J.A. Gulland (Eds.), *Fish Population Dynamics.* (pp. 216-239). Chichester, N.Y.: J. Wiley & Sons.

Garrod, D.J. (1982). *Stock and Recruitment-again.* (Fish. Res. Tech. Rep. No. 68). MAFF Direct. Fish. Res. (Lowestoft).

Garrod, D.J. (1988). The North Atlantic Cod: Fisheries and Management to 1986. In J.A. Gulland (Eds.), *Fish Population Dynamics* (pp. 185-218). Chichester, N.Y.: J. Wiley & Sons.

Garrod, D.J. and Colebrook, J.M. (1978). Biological Effects of Variability in the North Atlantic Ocean. *Rapp. Procès-Verb. Cons. int. Explor. Mer., 173*, 128-144.

Gean, G. and Nayar, M. (1988). ITQs in the Southern blue fin tuna fishery. *Marine Resource Economics, 5(4)*, 365-387.

Getz, W.H. (1984). Production models for nonlinear stochastic age structured fisheries. *Math. Biosci., 69*, 11-30.

Getz, W.M., Francis, R.C. and Swartzman, G.L. (1987). On managing variable marine fisheries. *Can. J. Fish. Aquat. Sci., 44*, 1370-75.

Glantz, M.H. and Thompson, J.D. (Ed.). (1981). *Resource Management and Environmental Uncertainty*. N.Y.: J. Wiley.

Goodland, R., Ledec, G. and Webb, M. (1989). Meeting Environmental Concerns Caused by Common-Property Mismanagement. In Berkes Fikret

(Eds.), *Economic Development Projects in Common Property Resources, Ecology and Community-based Sustainable Development* (pp. 148-167). London: Belhaven Press.

Goodyear, C.P. and Cristensen, S.W. (1984). On the ability to detect the influence of spawning stock on recruitment. *N. Am. J. Fish. Mgt., 4*, 186-193.

Gordon, H. (1954). The economic theory of the common property resource: the fishery. *J. Pol. Econ, 62(2)*, 124-142.

Gordon, H.S. (1956). Obstacles to Agreement of Control in the Fishing Industry. In R. Turvey and J.W. Wiseman (Eds.), *The Economics of Fisheries*. (pp. 65-72). Rome: FAO.

Gulland, J.A. (1982). Why do fish numbers vary? *Theor. Pop. Biol., 97*, 69-75.

Gulland, J.A. and Garcia, S. (1984). Observed Patterns in Multispecies Fisheries. In R.M. May (Eds.), *Exploitation of Marine Communities*. Springer-Verlag.

Gundmundsson, G. (1986). Statistical consideration in the analysis of catch-at-age observation. *J. Cons. Int. Explor. Mer., 93*, 83-90.

Hanley, N.D. (1989). *Contingent Valuation as a Method for Valuing Changes in Environmental Service Flows* (Discussion Paper in Economics No. 89/8). Dep. of Economics, Univ. of Stirling.

Hausman, J.A. (1979). Individual Discount Rates and the Purchase and Utilization of Energy Using Durables. *J. Econ., 10*, 33-54.

Hilborn, R. (1979). Comparison of fisheries control systems that utilize catch and effort data. *J. Fish. Res. Bd. Can., 36*, 1477-1489.

Hilborn, R. (1985). Apparent stock recruitment relationships in mixed stock fisheries. *Can. J. Fish. Aquat. Sci., 42*, 718-723.

Hilborn, R. and Walters, C.J. (1992). *Quantitative Fisheries Stock Assessment Choice, Dynamics and Uncertainty*. N.Y., London: Routledge, Chapman & Hall, Inc.

Hilden, M. (1988). Errors of perception in stock and recruitment studies due to wrong choices of natural mortality rate in VPA. *J. Cons. Int. Explor. Mer., 44*, 982-988.

Horwood, J. (1991). An approach to better management: the North Sea haddock. *J. Cons. int. Explor. Mer., 47*, 318-322.

Horwood, J.W., Bannister, R.C.A. and Howlett, G.L. (1986). Comparative fecundity of North Sea plaice (*Pleuronectes platessa* L.). *Proc. R. Soc. B, 228*, 401-431.

Horwood, J.W. and Shepherd, J.G. (1981). The sensitivity of age structured populations to environmental variability. *Math. Biosci., 57*, 59-82.

Intriligator, M.D. (1971). *Mathematical Optimization and Economic Theory.* Englewood Cliffs, N.J.: Prentice-Hall.

Jentoft, S. (1985). Models of fishery development: the cooperative approach. *Marine Policy, 9,* 322-331.

Jones, R. (1984). *Assessing the Effects of Changes in Exploitation Pattern Using Length Composition Data (with notes on VPA and cohort analysis).* (Fish. Tech. Paper No. 256). FAO (Rome).

Kamien, M.I. and Schwartz, N.L. (1981). *Dynamic Optimization, the Calculus of Variations and Optimal Control in Economics and Management.* Maurice Willinson.

Kennedy, J.O.S. (1989). *The Determination of the Optimal Exploitation Pattern of Western Mackerel Stocks.* (Seafish Report No. 3001). Sea Fish Industry Authority.

Kennedy, J.O.S. (1991). I.T.Q.s: The Australian experience. In *EAFE Annual Conference*, 12 (pp. 1-17). Dublin:

Keyfitz, N. (1968). *Introduction to the Mathematics of Population.* Reading, Mass.: Addison-Wesley.

Kirkley, J.E. and Strand, I.E. (1988). The technology and management of multi-species fisheries. *Appl. Econom., 20,* 1279-1292.

Koslow, J.A. (1989). Managing nonrandomly varying fisheries. *Can. J. Fish. Aquat. Sci., 46,* 1302-8.

Lai, H.L. and Gallucci, V.F. (1988). Effects of parameter variability on length-cohort analysis. *J. Cons. Int. Explor. Mer, 45,* 82-91.

Laidler, D. and Estrin, S. (1989). *Introduction to Microeconomics.* (3rd ed.). Philip Allan.

LaPointe, M.F., Peterman, R.M. and MacCall, A.D. (1989). Trends in fishing mortality rate along with errors in natural mortality rate can cause spurious trends in fish stock abundances estimated by virtual population analysisis (VPA). *Can. J. Fish. Aquat. Sci., 46,* 2129-2139.

Larkin, P.A. (1977). An Epitaph for the concept of Maximum Sustained Yield. *Trans. Am. Fish. Soc., 106(1),* 1-11.

Larkin, P.A. (1988). Pacific Salmon. In J.A. Gulland (Eds.), *Fish Population Dynamics.* (pp. 153-183). Chichester, N.Y.: J. Wiley & Sons.

Laurec, A. and Shepherd, J.G. (1983). On the analysis of catch and effort data. *J. Cons. Int. Explor. Mer., 41,* 81-84.

Ledec, G. and Goodland, R. (1988). *Wldlands: their Protection and Management in Economic Development.* No. World Bank, (Washington).

Lee, E.B. and Markus, L. (1968). *Foundations of Optimal Control Theory.* N.Y.: J. Wiley & sons.

Leslie, P.H. (1948). Some further notes on the use of matrices in population mathematics. *Biometrics, 35*, 213-245.

Loomis, J.B. and Walsh, R.G. (1986). Assessing wildlife and environmental values in cost-benefit analysis: state of the art. *J. Envir. Manag., 22*, 125-131.

Lovejoy, W.S. (1986). Bounds on the optimal age-capture for stochastic, age-structured fisheries. *Can. J. Fish. Aquat. Sci., 43*, 101-107.

Ludwig, D and Walters, C.J. (1985). Are age structured models appropriate for catch-effort data? *Can. J. Fish. Aquat. Sci., 42*, 1066-1072.

Ludwig, D and Walters, C.J. (1989). A robust method for parameter estimation from catch and effort data. *Can. J. Fish. Aquat. Sci., 46*, 137-144.

May, R. M. (1975). Biological populations obeying difference equations: stable points, stable cycles and chaos. *J. Theor. Biol., 51*, 511-524.

McCall, A.D. (1980). The consequences of cannibalism in the stock-recruitment relationship of planktivorous pelagic fishes such as Engraulis. *IOC Workshop, 28*.

McKellar, N. (1990). *The ITQ Debate* (Occasional papers No. Sea Fish Industry Authority.

Meany, T.F. (1978). Restricted Entry in Australian Fisheries. In B.R. Rettig and J.J.C. Ginter (Eds.), *Limited Entry as a Fishery Management Tool* (pp. 391-415). Washington (D.C.): Univ. Washington Press.

Megrey, B.A. (1989). Review and comparison of age-structured stock assessment models from theoretical and applied points of view. *Am. Fish. Soc. Symp., 6*, 8-48.

Miller, D.L. (1989). The Evolution of Mexico's Spiny Lobster Fishery. In F. Berkes (Eds.), *Economic Development Projects in Common Property Resources, Ecology and Community-based Sustainable Development* (pp. 185-198). London: Belhaven Press.

Miller, R.E. (1970). *Dynamic Optimization and Economic Applications*. N.Y.: McGraw-Hill Inc.

Monaghan, P., Uttley, J.D., Burns, M.D., Thaine, C. and Blackwood, J. (1989). The Relationship Between Food Supply, Reproductive Effort and Breeding Success in Arctic Terns *Sterna Paradisea*. *J. Anim. Ecol., 58*, 261-274.

Mueller, J.J. and Wang, D.H. (1981). Economic Analysis in Fishery Management Plans. In Anderson L.G. (Eds.), *Economic Analysis for Fisheries Management Plans* London, N.Y.: Ann Arbor.

Munro, G.R. and Scott, A.D. (1985). The Economics of Fisheries Management. In A.V. A.V. Kneese and J.L. Sweeney (Eds.), *Handbook of*

Natural Resource and Energy Economics. (pp. 623-676). North-Holland: Elsevier Science Publishers B.V.

Murphy, G.I. (1973). *Clupeoid Fishes Under Explotation with Special Reference to the Peruvian Anchovy.* (Tech. Rep. No. 30). Hawaii Inst. of Marine Biology.

Neary, J.P. and Roberts, K.W.S. (1980). The theory of household behaviour under rationing. *Europ. Econ. Rev., 13,* 25-42

Neher, P.A. (1990). *Natural Resource Economics Conservation and Exploitation.* N.Y., Melbourne: Cambridge Univ. Press.

Noel, E. (1989). The Single European Act: Meaning and Prospective. *Contemporary European Affairs, 1,* 93-108.

O'Neil, R.V. (1973). Error analysis of ecological models Deciduous Forest Biome Memo Report, 71,15.

Page, T. (1976). *Conservation and Economic Efficiency.* Baltimore: J. Hopkins, Univ. Press.

Pauly, D. and Murphy, G.I. (Ed.). (1982). *Theory and Management of Tropical Fisheries.* Manila.

Pearce, D.W. (Ed.). (1986). *The Mit Dictionary of Modern Economics.* (3rd ed.). Cambridge, Massachusetts: Mit Press.

Pearce, D.W. and Turner, R.K. (1990). *Economics of Natural Resources and the Environment.* Harvester Wheatsheaf.

Pearse, P.H. (1979). Symposium on Policies for Economic Rationalization of Commercial Fisheries. *J. Fish. Res. Board. Can., 36,* 711-866.

Pearse, P.H. (1980). *Regulation of Fishing Effort* (Fish. Tech. Paper No. 197). FAO.

Plourde, C.G.} (1971). Exploitation of Common-property Replenishible Resources. *West. Econ., 9,* 256-266.

Plourde, G.C. (1970). A Simple Model of Replenishible Resource Exploitation. *Amer. Econ. Rev., 60,* 518-522.

Pontryagin, L.S., Boltjanskii, V.S., Gamkrelidze, R.V. and Mishchenko (1962). *The Mathematical Theory of Optimal Processes.* N.Y.: Wiley.

Pope, J.G. (1992). No. 22. STCF of the European Commission.

Pope, J.G. and Knight, B.J. (1982). Simple Models of Predation in Multi-Age Multispecies Fisheries for Considering the Estimation of Fishing Mortality and its Effects. In M.C. Mercer (Eds.), *Multispecies Approaches to Fisheries Management Advice.* (pp. 64-69). Canadian Special Publications on Fish. Aquat. Sci.

Pope, J.G. and Shepherd, J.G. (1985). A comparison of the performance of various methods for tuning VPAs using effort data. *J. Cons. Int. Explor. Mer, 42,* 129-151.

Pope, J.G. and Stokes, T.K. (1987). The use of multiplicative models for separable VPA integrated analysis, and the general VPA tuning problem. *Trans. Am. Fish. Soc., 116.*

Quinn, T.J., Deriso, R.B. and Hoag, S.H. (1985). *Methods of Population Assessment of Pacific Halibut.* No. 72. IPHC Scientific Report.

Quirk, J. and Smith, V. (1970). Dynamic Economic Models of Fishing. In A. Scott (Eds.), *Economics of Fisheries Management: a Symposium.* J. Wiley & sons.

Ramsey, F.P. (1928). Mathematical Theory of Saving. *Econ. J.(38),* 543-559.

Reed, W.J. (1983). Recruitment variability and age structure in harvested animal populations. *Math. Biosci., 65,* 239-268.

Rettig, B.R. and Ginter, J.J. (Ed.). (1978). *Limited Entry as a Fishery Management Tool.* Washington (D.C.): Univ. V Press.

Ricker, W.E. (1954). Stock and Recruitment. *J. Fish. Res. Board Can., 11,* 559-563.

Ricker, W.E. (1973). Two mechanisms that make it impossible to maintain peak period yields from Pacific Salmon and other fishes. *J. Fish. Res. Board Can., 30,* 1275-1286.

Rivard, D. and Foy, M.G. (1987). An analysis of error in catch projections for Canadian Atlantic fish stocks. *Can. J. Fish. Aquat. Sci., 44,* 967-981.

Ruddle, K. (1989). Solving the Common-Property Dilemma: Village Fisheries Rights in Japanese Coastal Waters. In F. Berkes (Eds.), *Economic Development Projects in Common Property Resources, Ecology and Community-based Sustainable Development* (pp. 168-184). London: Belhaven Press.

Sampson, D.B. (1988). The stability of virtual population analysis in cohort size estimates. *J. Cons. Int. Explor. Mer., 44,* 135-142.

Sampson, D.B. (1991). Fishing tactics and fish abundance and their influence on catch rates. *ICES J. Mar. Sci., 48,* 291-301.

Samuelson, P.A. (1965). A catenary turnpike theorem involving consumption and the golden rule. *Amer. Econ. Rev., 55,* 486-496.

Santander, H., Alheit, J., McColl, A.D. and Alaino, A. (1983). Egg mortality of the Peruvian anchovy (Engraulis ringens) caused by cannibalism and predation by sardines (*Sardinops sagax*). *FAO Fish. Rep., 291*(3).

Saville, A and Bailey, R.S. (1980). The assessment and management of the herring stocks in the North Sea and to the west of Scotland. *Rapp. P.-V. Reun. Cons. Int. Explor. Mer., 177,* 112-142.

Schaefer, M. (1954). Some aspects of the dynamics of populations important to the mangement of the commercial marine fisheries. *Bull. Inter-Amer. Tropical Tuna Comm., 1,* 25-36.

Schaefer, M. (1957). Some Considerations of the population dynamics and economics in relation to the management of the commercial marine fisheries. *J. Fish. Res. Board. Can., 14*, 669-681.

Schaefer, M. (1959). Biological and economic aspects of the management of the commercial marine fisheries. *Trans. Amer. Fish. Soc., 88*, 100-104.

Schnute, J. and Richards, L.J. (1990). A unified approach to the analysis of fish growth, maturity and survivorship data. *Can. J. Fish. Aq. Sci., 47*, 24-40.

Scott, A.D. (1955). The fishery: the objectives of sole-ownership. *J. Pol. Econ., 63*, 116-124.

Segerson, K. and Squires D. (1993). Capacity Utilization Under Regulatory Constraints. *Rev. Econ. Statist., 75(1)* February 1993, 76-85.

Shelton, P.A. and Armstrong, M.J. (1983). Variations in parent stock of pilchard and anchovy populations in the Southern Bengouela system. *FAO Fish. Rep., 293(3)*, 1113-1132.

Shepherd, J.G. (1988). Fish Stock Assessments and their Data Requirements. In J.A. Gulland (Eds.), *Fish Population Dynamics* (pp. 35-62). Chichester, N.Y.: J. Wiley & Sons Ltd.

Sims, S.E. (1984). An analysis of the effect of errors in the natural mortality rate on stock-size estimates using VPA (cohort analysis). *J. Cons. Int. Explor. Mer., 41*, 149-153.

Sinclair, M., Anthony, V.C., Iles, T.D. and O'Boyle, R.N. (1985). Stock assessment problems in Atlantic herring (Clupea harengus) in the Northwest Atlantic. *Can. J. Fish. Aquat. Sci., 42*, 888-898.

Sissenwine, M.P. (1984). Why Do Fish Populations Vary? In R.M. May (Eds.), *Exploitation of Marine Communities.* (pp. 59-94). Spinger-Verlag.

Slaymaker J.E. (1992) "The role of producers organizations in the fisheries management: the case of South West England" in proceedings of the *IV Annual Conference of the European Association of fisheries Economists*, 1992, pp.37-46.

Smith, A.H. and Berkes, F. (1991). Solutions to the "tragedy of the Commons": Sea-urchin Management in St. Lucia, West Indies. *Environmental Conservation, 18(2)*, 131-136.

Smith, T.D. (1988). Stock Assessment Methods: the First Fifty Years. In J.A. Gulland (Eds.), *Population Dyanmics.* (pp. 1-33). Chichester, N.Y.: J. Willey & Sons Ltd.

Smith, V.L. (1969). On models of commercial fishing. *J. Pol. Econ., 77*, 181-198.

Solow, R.M. (1974). The economics of resources or the resources of economics. *Amer. Econ. Rev., 64*, 1-14.

Spence, M., Starrett, D. and Rev., Int. Econ. (1975). Most rapid approach paths in accumulation problems. *Int. Econ. Rev., 16(2)*, 388-403.

Stevens, T.H., Echeverria, J., Glass, R.J., Hager, T. and Thomas, A. (1991). Measuring the existence value of wildlife: what do CVM estimates really show?". *Land Econ., 67*, 490-500.

Troadec, J.P., Clark, W.G. and Gulland, J.A. (1980). A review of some pelagic fish stocks in other areas. *Rapp. Procès-Verb. Cons. int. Explor. Mer., 177*, 252-277.

Tucker, C.E. and McKellar, N.B. (1993). The Value Of Not Catching Fish (or The Shadow Value of Fish Stocks) Seminar presented in E.A.F.E. Vth Conference Brussels, Belgium.

Ulltang, O. (1976). Catch per unit of effort in the Norwegian purse seine fishery for Atlanto-Scandinavian herring. *FAO Fish. Tech. Rap., 155*, 91-101.

Vance, R.R. (1973). On reproductive strategies in marine benthic invertebrates. *Am. Nat., 107*, 339-352.

Vetter, E.F. (1988). Estimation of natural mortality in fish stocks: a review. *Fish. Bull., 86*, 25-42.

Walters, C.J. (1986). *Adaptive Management of Renewable Resources*. N.Y.: Collier Macmillan.

Wise, M. (1984). *The CFP of the Europen Community* London, N.Y.: Methuen & Co Ltd.

Whitehead, J.C. (1993). Total economic values for coastal and marine wildlife: specification, validity and valuation issues. *Marine Resource Economics, 8*, 119-132.

Young T. (1984) "Market Support Arrangements for Fish: the Withdrawal Price Scheme", *Sea Fish Industry Authority* Occasional Paper Series No. 1.

Appendices

Appendix 1 **Total landings of species for human consumption and industrial species**

Average quantities for years 1988-1989 (x1000tons)

Member States	Industrial species	Species for human consumption
Belgium	0.017	31
Germany	14.6	140.4
Denmark	1299.5	590.5
Greece	1.2	146.3
Spain	42.2	949.3
France	3.7	689.3
Ireland	22.4	204.6
Italy	3.1	372.4
Netherlands	35.2	375
U.K.	67.2	714.3
Portugal	31	309.5

sources: Commission of the E.C., Eurostat Statistical Office, "Fisheries:Yearly Statistics 1991", Luxembourg Office for Official Publications of the E.C., 1992 and D.G. XIV, XIV-A-3, 30.6.1993

Appendix 2   Data for North Sea demersal fisheries

COD

Years	1981	1982	1983	1984	1985	1986	1987	1988	1989	1990	1991	1992
Recomended TAC	220	235	220	215	259	130	200	148	124	113	92	92
Agreed TAC	220	235	240	215	250	170	175	160	124	105	100	100
Catch	301	273	234	205	193	163	175	150	116	105	86	
SSB	173	168	135	116	107	95	86	79	73	62	56	51

HADDOCK

Years	1981	1982	1983	1984	1985	1986	1987	1988	1989	1990	1991	1992
Recomended TAC	140	200	170	172	209	239	120	185	68	50	48	48
Agreed TAC	140	180	181	170	207	230	140	185	68	50	50	60
Catch	204	242	238	213	251	220	172	171	104	87	90	
SSB	228	285	241	190	231	213	150	149	119	71	55	105

SAITHE

Years	1981	1982	1983	1984	1985	1986	1987	1988	1989	1990	1991	1992
Recomended TAC	127	100	131	160	195	195	198	156	170	120	125	102
Agreed TAC	127	125	158	180	200	240	173	165	170	120	125	110
Catch	136	173	173	195	200	164	149	105	92	88	96	
SSB	224	188	196	138	107	100	102	82	71	66	56	68

WHITING

Years	1981	1982	1983	1984	1985	1986	1987	1988	1989	1990	1991	1992
Recomended TAC	120	200	125	102	118	135	127	134	115	130	135	135
Agreed TAC	150	170	170	149	160	135	135	120	115	125	141	135
Catch	191	150	151	135	97	154	132	127	118	147	117	
SSB	451	349	303	247	242	263	273	267	243	264	291	305

PLAICE

Years	1981	1982	1983	1984	1985	1986	1987	1988	1989	1990	1991	1992
Recomended TAC	105	140	164	150	130	160	120	150	175	171	169	169
Agreed TAC	105	140	164	182	200	180	150	175	185	180	175	175
Catch	140	155	144	156	160	165	160	162	170	168	154	
SSB	296	285	310	307	345	350	342	377	433	402	346	385

SOLE

Years	1981	1982	1983	1984	1985	1986	1987	1988	1989	1990	1991	1992
Recomended TAC	15	15	15	14	15	12	11	11	14	25	27	21
Agreed TAC	15	21	20	20	22	20	14	14	14	25	27	25
Catch	15.4	21.6	24.9	26.6	24.2	18.2	17.4	21.6	21.8	35.1	38.0	
SSB	25.3	35.4	41.5	44.4	43.0	37.0	32.0	43.0	38.0	94.0	80.0	74.0

sources: ICES Cooperative research reports. Reports of the ICES Advisory Committee on Fisheries Management. Volumes: No193, part 1, chapter 3.5, February 1993; No179, part 1, chapter 3.5, February 1992; No161, part 1, chapter 3.5, February 1989.

274

Appendix 3   **The average landing price per tonne for fishery products in ten E.C. States in 1990**

Member States	ECUs
Germany	740.260
Belgium	2258.065
Denmark	295.939
France	1662.500
Greece	3367.347
Ireland	459.893
Italy	3727.003
U.K.	901.800
Spain	1941.788

source: D.G. fisheries XIV-A-3.(30.6.1993)

Community expenditure under the intervention scheme of the Common Market Organization for fishery products

Years	million ECUs
1983	17.15
1984	14.54
1985	18.41
1986	17.23
1987	17.38
1988	47.05
1989	23.93
1990	23.60
1991	29.00
1992	27.00

note: Data1983-1985 for 10 Member States, data 1986-1992 for 12 Member States

source: SEC(91) 2288 final (18.12.1991) and O.J. (8.2.1993) L31

Average landing price per tonne of all fisheries products in E.C.

Years	ECUs
1973	341.913
1974	346.739
1975	358.322
1976	397.094
1977	467.592
1978	465.630
1979	496.774
1980	498.709
1981	596.506
1982	622.653
1983	689.981
1984	734.189
1985	731.502
1986	1005.401
1987	1139.018
1988	1122.251
1989	1098.646
1990	1244.947

note: Data 1973-1980 for 9 Member States, data1981-1985 for 10 Member States, data 1986-1992 for 12 Member States

source: D.G. XIV, XIV-A-3 (30.6.1993)

Appendix 6 Supply of fish in the Community

Years	Landings (1000 tonnes)	Imports (1000 tonnes)	Exports (1000 tonnes)	Total Supply (1000 tonnes)
1976	4543	1687	501	5729
1977	4397	1805	525	5677
1978	4437	2131	806	5762
1979	4340	2375	1018	5697
1980	4646	2346	1001	5991
1981	4694	2205	1068	5831
1982	4900	2609	1217	6292
1983	4677	2477	1315	5839
1984	4638	2731	1266	6103
1985	4642	3233	1303	6572
1986	6850	3322	1306	8866
1987	5172	3613	1237	7548
1988	5456	3548	1239	7765
1989	5128	3991	1328	7791
1990	4440	4160	1220	7380

note: Data 1976-1980 for 9 Member States, data1981-1985 for 10 Member States, data 1986-1992 for 12 Member States

sources: D.G. XIV, XIV-A-3, 30.6.1993, SEC(91) 2288 final, 18.12.1991 and Review of Fisheries in OECD Member Countries, Paris, 1993

Consumption flow of fish products per capita in the E.C.

Years	Consumption per capita (kg) per year
1970	16.47
1971	17.4
1972	19.92
1973	16.79
1974	18.34
1975	18.35
1976	18.94
1977	18.99
1978	19.82
1979	18.67
1980	19.08
1981	18.22
1982	17.6
1983	17.67
1984	17.14
1985	17.44
1986	19.59
1987	20.1
1988	20.94
1989	16.9
1990	17.81

note: Data 1970-1972 for six Member States, data 1973-1980 for 9 Member States, data 1981-1985 for 10 Member States, data 1986-1992 for 12 Member States

Data for fish consumption must be treated sceptically particularly since 1986 as there is some inconsistency among different sources, e.g. FAO Yearbooks fisheries statistics, Eurostat (1A) basic statistics and Agriculture statistical yearbooks

source: D.G. fisheries, XIV-A-3

279

Appendix 8   Annual fish landing price (£) per tonne (mean) for six species landed by U.K. fishing vessels in Scottish ports. S.D.: Standard Deviation of monthly prices.

Years	cod mean	cod S.D.	haddock mean	haddock S.D.	saithe mean	saithe S.D.	whiting mean	whiting S.D.	herring mean	herring S.D.	Mackerel mean	Mackerel S.D.
1975	264.5	47.984	216.417	27.953	103.5	37.805	143.667	16.946	108.417	29.159	65.364	54.478
1976	375.083	81.169	260.25	35.397	175.917	62.236	185.25	27.05	145.5	48.998	110.167	70.997
1977	551.167	50.668	384.75	44.884	281.167	46.307	286.833	30.936	345.75	106.773	142.417	83.04
1978	571.5	64.189	485.917	68.814	292.75	59.629	301.833	31.054	429.333	101.529	166.182	84.133
1979	627.833	71.222	504.75	58.205	354.333	79.434	334.25	45.279	574.09	88.85	184.7	133.516
1980	612.917	72.898	424.75	66.101	350.5	79.193	314.167	47.263	413.75	46.351	134.09	100.73
1981	564.667	79.314	394.417	71.917	278.083	51.205	307.167	54.416	295.92	163.99	140.25	72.946
1982	649.667	49.325	391.833	56.886	294.58	47.98	302.75	56.7	209.25	133.93	149.375	142.75
1983	699.5	107.985	484.75	81.978	303.417	73.335	330.417	63.902	186.75	125.42	256.083	169.381
1984	720	63.5	611.333	74.657	240.417	57.578	411	49.216	137.636	53.886	156.728	134.326
1985	758.75	55.057	546.75	136.746	283.167	55.816	423.167	83.346	136.727	49.64	146.727	72.02
1986	905.333	67.334	614.417	96.545	375.417	71.657	481.833	63.882	128	51.602	153.25	108.996
1987	939.833	86.636	764.333	33.268	467.583	59.95	517.5	92.116	144	81.459	165.364	87.459
1988	988.417	112.581	720.167	132.7	412.25	104.385	491.25	110.589	178.667	90.279	175.91	96.457
1989	1061.583	137.047	874.75	93.729	424.083	91.47	583.5	67.772	126.818	41.426	213	123.83
1990	1298.25	138.558	1228.083	88.375	483.333	126.568	741.583	157.837	192.083	126.616	281.667	165.215

source: Own calculations from data in Scottish Sea Fisheries Statistical Tables, extracts from 1975-1990

Appendix 9   **Price per tonne of horse mackerel in seven E.C. states**

Value, in U.S. $, by national landings in domestic ports of seven Member States, per tonne, for Horse Mackerel in 1990

Belgium	324.177
France	761.813
Germany	314.224
Ireland	231.405
Italy	1339.592
Portugal	1491.407
U.K.	171.313

source: Review of fisheries in OECD Member Countries, Paris 1993

Appendix 10   **Flow of Community trade in fishery products during 1983-1992**

Years	Imports million ECUs	Exports million ECUs	Trade balance million ECUs
1983	3321	1110	-2211
1984	3681	1217	-2464
1985	4132	1371	-2761
1986	4593	1266	-3327
1987	5294	1363	-3931
1988	5917	1372	-4545
1989	6390	1462	-4928
1990	6925	1352	-5573
1991	7698	1471	-6227
1992	7655	1420	-6235

note: Data1983-1985 for 10 Member States, data 1985-1992 for 12 Member States

source: D.G. Fisheries, XIV-A-3, situation at 30.VI.1993

Evolution of E.C. expenditure on fisheries agreements with third countries

Years	Million ECUs
1983	10.1
1984	12.9
1985	29.5
1986	29.1
1987	59.1
1988	110.3
1989	116.7
1990	173.1
1991	171.4
1992	101.0
1993	156.0

sources: Yearly Official Journals on Final adoption of the general budget of the European Communities for the years 1983-1993

Appendix 12   E.C. annual expenditure on the construction and modernization of the fishing fleet

Years	construction and modernization (Million ECU)	Shipbuilding (Million ECU)	Modernization (Million ECU)	Withdrawals (Million ECU)
1971	1.47			0.14
1972	7.47			0.77
1973	7.47			0.62
1974	16.62			0.51
1975	12.15			6.50
1976	13.99			6.71
1977	2.89			4.61
1978	8.47			6.85
1979	15.02			8.50
1980	11.66			11.56
1981	13.54			15.39
1982	20.62			8.22
1983	28.3	21.3	7	6.85
1984	49.5	39.1	10.4	6.85
1985	62	46.8	15.2	6.85
1986	56.7	38.7	18	6.85
1987	73.2	64	9	26.2
1988	28	8.7	19.3	31.2
1989	83.8	63.5	20.3	25.7
1990	70.4	44.2	26.2	49.8
1991	81.66	47.24	34.42	34.1

note: Data 1971-1972 for 6 Member States, data 1973-1980 for 9 Member States, data 1981-1985 for 10 Member States, data 1986-1992 for 12 Member States

sources: D.G. XIV, XIV-A-3 30.6.1993, SEC(91) 2288(final) 18.12.1991, O.J. (15.12.1992) C330/107, and Court of Auditors, Special report No 3/93

Appendix 13   Trend of Community's commitment appropriations for fisheries during 1983 to 1993

Years	C.O.M. (million ECUs)	Structural Policy (million ECUs)	Fisheries agreements (million ECUs)	Control (million ECUs)	Research (million ECUs)	Total expenditure (million ECUs)
1983	25.45	53.55	10.1	0.3	0.6	90
1984	14.59	78.45	12.9	0.3	0.6	106.84
1985	18.46	104.35	29.5	0.3	1.3	153.91
1986	17.23	117.95	29.1	0.3	0.8	165.38
1987	17.45	151.3	59.1	0.5	0.5	228.85
1988	47.26	143.2	110.3	11	4.3	316.06
1989	23.96	210.5	116.7	1	11	363.16
1990	23.61	232.7	173.1	15.8	10.6	455.81
1991	26.63	289.9	171.4	24	13.1	525.03
1992	29.60	325.3	101.1	23.7	13.1	492.7
1993	28.00	390.5	156	25	19	618.5

note: Data1983-1985 for 10 Member States, data 1986-1993 for 12 Member States.
C.O.M.: Common Organization of the Market

sources: SEC(91) 2288 (18.12.1991) and General Budget of EC in O.J. (8.2.1993) L31.

Community's total commitments and payments for fisheries and their rate of utilization for each year during 1987 to 1993

Years	Total expenditure (million ECUs)	Payments (million ECUs)	Utilization rate (%)
1987	228.85	157.8	69
1988	316.06	260	82.3
1989	363.16	261.9	72.1
1990	455.81	325.6	71.4
1991	525.03	363	69.1
1992	492.70	342.28	69.5
1993	618.50	415	67.1

source: Data from O.J. (15.2.1993) C330 and General Budget of the E.C. for the financial year 1993 in O.J. (8.2.1993) L31